MATHEMATICS,

SCIENCE, AND

POSTCLASSICAL

THEORY

MATHEMATICS, SCIENCE, AND POSTCLASSICAL THEORY

Barbara Herrnstein Smith and

Arkady Plotnitsky, Editors

Duke University Press   Durham & London 1997

© 1997 Duke University Press
All rights reserved
Printed in the United States of America on acid-free paper ♾
Typeset in Scala by Tseng Information Systems, Inc.
Library of Congress Cataloging-in-Publication Data appear
on the last printed page of this book.
The text of this book was originally published, without
the essay "Microdynamics of Incommensurability" by
Barbara Herrnstein Smith and without the index, as
Volume 94, Number 2 of the *South Atlantic Quarterly*.

Contents

MATHEMATICS,

SCIENCE, AND

POSTCLASSICAL

THEORY

Introduction: Networks and Symmetries,

Decidable and Undecidable

Barbara Herrnstein Smith

and Arkady Plotnitsky

T HE aim of this volume is to indicate the scope, vitality, and general interest of recent work in and on mathematics and science, especially where it engages themes and issues that have also been central to contemporary cultural and literary theory and to a number of turns, recent and not so recent, in philosophy. While the term "postclassical theory" can be given a range of meanings in relation to more or less radical developments in all these fields, such as experimental and theoretical discoveries in quantum physics or significantly revisionary accounts of evolutionary dynamics in contemporary theoretical biology, its use here is intended primarily to evoke the various critical analyses and efforts at reconceptualization (again, more or less radical) that have emerged in the humanities and social sciences around a cluster of quite general but problematic concepts, notably, *knowledge, language, objectivity, truth, proof, reality,* and *representation,* and around such related issues as the dynamics of intellectual history, the project of foundationalist epistemology, and the distinctive (if they are distinctive) operations of mathematics and science.

Key figures here include, of course, Nietzsche, James, Peirce, Wittgenstein, Heidegger, Kuhn, Feyerabend, Foucault, Derrida, and Rorty, but also, as these essays indicate, Niels Bohr, Samuel Beckett . . . and Shakespeare. As they also indicate, a sense of the conceptual and practical (including technical) inadequacies of traditional formulations of such ideas is not confined to a handful of wayward Continental thinkers or local malcontents. More or less elaborately articulated critiques of all of them have emerged recurrently throughout the present century both from the diverse disciplines that study the operations of science, such as history, sociology, and philosophy, and from mathematics and the scientific disciplines themselves, including such new and hybrid fields as ecology, robotics, and neuroscience. These critiques proceed from, among other things, the dissatisfaction of practitioners in such disciplines who have found that familiar or classic accounts of knowledge, proof, truth, reality, and so forth do not cohere with empirical descriptions or mathematical analyses, cannot capture the complex dynamic processes of organic development and individual cognition, and require increasingly questionable interpretations of intellectual and scientific history.

The prominence given here to mathematics reflects the initial occasion of a number of these essays in a colloquium on Mathematics and Postclassical Theory sponsored in fall 1993 by Duke University's Center for Interdisciplinary Studies in Science and Cultural Theory. Since the Center's initial year of operation had focused on *biology*, it seemed a good idea to follow with a topic that indicated clearly, even dramatically, the range of disciplines that its activities were designed to embrace; and, in relation to biology, *mathematics* is often seen as occupying a position at the opposite end of a number of quite vivid scales: of ethereality and materiality, for example, or of durability and mutability, or even, as in the lingo of the artificial-intelligence and artificial-life communities, of dryness (as in the new silicon-based versions of those attributes) and wetness (as in their more familiar carbon-based versions). Quaint or otherwise questionable as these scales and metaphors may be, they are not without some interest here, for they reflect conceptual dualisms and more or less overt hierarchies that are the legacy — with important theoretical as well as institutional consequences — of ancient oppositions considered in several of these essays.

A second and more significant reason for the initial coupling of postclassical theory with mathematics was precisely the fact that the latter is so often invoked as an exception, prohibitive limit, or clear counterinstance to the more radical reaches of such theory: to post-Nietzschean epistemology, for example, or to post-Kuhnian history and sociology of science. Thus it is argued that the (supposed) solipsism, social determinism, or cultural relativism of these projects and approaches, along with their evident and (it is thought) exaggerated concern with language and rhetoric, is refuted by the manifest objectivity of mathematical entities and the manifest transcultural, transhistorical validity of mathematical knowledge. It is also argued that, in challenging these latter notions (i.e., objectivity and transcultural or transhistorical validity), postclassical theory (under whatever name is employed for purposes of castigation — "postmodernism," "deconstruction," "irrationalism," etc.) threatens to undermine the public confidence and professional self-confidence that secure the proper intellectual authority and technological benefits of mathematics and the sciences and ensure their continued pursuit. By contrast, although none of the essays here engages this popular "but 2 + 2 still = 4" move directly, those dealing with mathematics suggest that the relations — both historical and conceptual — between mathematics and postclassical theory (in the sense indicated above) are on the whole quite cordial and that, even where those relations are complex, they do not involve any wholesale refutations or underminings in either direction.

This is not to say that there is no bite to these essays (meaning now the entire group, not only those focused on mathematics) or to the conceptual and methodological developments they report and exemplify. The more radical aspects of these developments in relation to the cluster of concepts mentioned above (knowledge, truth, proof, reality, etc.) are evident in the sense of scandal that in many quarters still attends any conjunction of science, including mathematics, with any "post" in advance of . . . well, the postlapsarian. The fact that radical *theoretical* challenges in any field are also likely to involve, sooner or later, challenges to its instituted structures and practices more generally is also evident in the frequent and not altogether inaccurate association of such conjunctions with threats to established order, academic and otherwise. Some implications of these theoretical and methodological developments for issues central to critical cultural studies (for example, the invocation of a monolithically and otherwise dubiously conceived "science" either to sustain entrenched traditional ideologies or to lend credibility to otherwise highly questionable social analyses and political programs) are suggested by a number of the essays here. The spelling out of such implications is not, however, the primary motive of any of them or, for better or worse, of the collection as such.

Nonspecialist readers glancing at these pages may find some of the idioms displayed — abstract, numerical, tabular, diagrammatical — initially formidable or alien. The degree of technical specificity here is, in fact, probably greater than that commonly met in publications on such topics addressed to a general readership. It is our belief, however, as it was the aim of those who contributed to and worked on this volume, that a substantive engagement with these essays does/should not require either a specialized knowledge of mathematics or extensive familiarity with any of the specific fields — physics, sociology, economics, biology, and so forth — they represent. On the contrary, readers should find that this collection offers a genial introduction to a range of important contemporary areas of study and related issues and approaches. They will certainly find the contributors to be, for the most part, an unusually imaginative, witty, and chatty lot (and, *caveat lector,* not above a prank or two).

The general liveliness exhibited in these pieces as well as the ease with which individual contributors move from Frege to Beckett or from Darwin to Rube Goldberg should actually be no surprise. Although contemporary relations between the sciences and fields such as literary and cultural studies are vexed by a conceptual history of dualistic models and an institutional history of disciplinary segregation, the fundamental arbitrariness of all conceptual models and the instability and reconfigurability of all disciplinary divisions can be seen as both an implicit theme of these essays and a moral to be drawn from the achievements they report and embody. Only a very confined — fundamentally theological, perhaps — notion of the human would represent the humanities as standing in some essentially adversarial relation to mathematics or to the scientific disciplines more generally. And only a very confined — fundamentally metaphysical, it appears — notion of science would represent its domains and pursuits, including mathematics, as *other* than social, cultural, discursive, human, and, as Brian Rotman insists in the opening essay here, corporeal and therefore mortal.

Allusion was made above to common conceptions of the exemplary objectivity of mathematical knowledge. Of course, questions concerning the nature and reality of mathematical objects (numbers, geometrical figures, quaternions, Hilbert spaces, etc.), considered in a number of essays here from a largely postclassical perspective, have also been raised throughout the history of classical thought. Classical questioning, however, while not without significant theoretical and even, in a sense, postclassical implications, has been aimed for the most part (there are some exceptions) at solidifying or increasing rather than disturbing or diminishing the power of classical ideas and values. From Parmenides and Plato to Husserl and Heidegger and, via Kant and Hegel, beyond, whenever mathematical knowledge has been found wanting (as unable, for example, to address issues of ethics or morals), it is usually because mathematics is seen as incapable of reaching the kind of knowledge — that is, truth — thought to be available to philosophy and, in most cases, to philosophy alone. In other words, throughout the history of philosophy and in classical thought more generally, mathematics has been seen either as the example par excellence of the classical ideal of knowledge or as insufficient in just that respect.

It is clear, however, that—even leaving aside the ongoing debate opened by Gödel's theorem—many classical assumptions concerning mathematical knowledge and its historical, social, and cultural conditions have come under increasing scrutiny, sometimes by leading practitioners in the field. As early as 1830, Evariste Galois, considered by some the greatest mathematical genius who ever lived (even though and perhaps because he was killed in a duel at the age of twenty-one), offered what we would now consider striking insights into the psychological, social, and indeed political constitution and functioning of mathematics. But then Galois, one of the greatest mathematical revolutionaries ever, was also a political revolutionary, which fact may well have cost him and mathematics his life. In any case, it appears that mathematics and politics have more junctures than one would expect, given classical conceptions of either. Some of these junctures are familiar, others are strange, and still others are both, simultaneously. They are, in short, in a specific sense discussed below, postclassical.

Another point should, however, be stressed here. No matter how radical the critique of the foundations—especially the philosophical foundations—of mathematical knowledge undertaken by postclassical theory, it is never a simple dismissal of those aspects of mathematics that make it exemplary from a classical perspective. As remarked above, the relations between postclassical theory and mathematics, as represented in this collection, are on the whole quite cordial. This does not, however, diminish the force of contemporary critiques of the nature of mathematical knowledge, nor does it diminish the pressures that continue to evoke them. If anything, the opposite is the case. Thus, while a quite pointed questioning of the metaphysical appurtenance of infinitist mathematics—that is, all mathematics insofar as the concept of infinity figures in it—emerges in Brian Rotman's essay, the contributions by John Vignaux Smyth and Arkady Plotnitsky suggest that postclassical theory and mathematics, including infinitist mathematics, are not only compatible, but may even be necessary to each other in certain areas of their functioning—for example, in science and technology (as suggested by Plotnitsky's essay) or, sometimes interactively, in such fields as literature, the arts, or culture as a whole (as suggested by Smyth's).

The configuration just described may be seen as a key characteristic of postclassical theory: that is, where familiar metaphysical assumptions are challenged, it is not always or necessarily in order to deny their relevance but, rather, to indicate either the limits of that relevance or the simultaneous relevance of opposite assumptions. Alternatively, one could say that such assumptions (or, strictly speaking, the propositions that articulate them), precisely in being metaphysical, are *undecidable*. The latter concept is crucial here and we return to it below.

While abstractness, in the sense of generality if not ethereality, could be seen as a defining attribute of theory, mathematics, and all scientific knowledge as such, several of the essays here are quite noticeably engaged by the sublunary and the concrete. For example, contemporary reconceptualizations of *materiality*, pursued along epistemological lines by Rotman and Plotnitsky, are addressed in what could be called socio-epistemological — or perhaps socio-antiepistemological — terms in the essays by Andrew Pickering, Michel Callon and John Law, and Malcolm Ashmore. One of the marks, in fact, of contemporary "social studies of science," a revisionist field represented here primarily by the latter three contributions, is the general willingness of those who work in it to retain a focus on the mundane details of scientific culture: the local, the ephemeral, and even in a sense the domestic (e.g., the laboratory telephones and pencils that acquire a certain prominence in Callon and Law's essay). What is significant here is that each of these three essays — and work in this field more generally — resists the tendency, characteristic of more classical accounts of science, to treat particular examples in such a way as to permit rapid ascent to an abstract, general, and thus presumptively transcendent level.

The reconceptualization of materiality reflected in the essays just mentioned may be seen as a logical extension of Bohr's insistence on the fundamental role of measuring instruments in quantum mechanics, as discussed here by Plotnitsky. It is also related to the material technologies of "virtual reality" and of "writing" as understood by Derrida, both of which are considered in their varied and sometimes surprising connections with mathematics in Rotman's essay ("Thinking Dia-Grams: Mathematics, Writing, and Virtual Reality"). Other significant aspects of materiality related to language emerge in Smyth's essay, again along Derridian (and here also de Manian) lines. Clearly, there are many ways in which one could map the interconnections among these essays as well as their individual connections to various aspects of postclassical theory. While the possibility of a fully determinate or decidable map — whether geographical, historical, or other — is itself made questionable by postclassical theory, that does not eliminate the possibility or indeed the necessity (as here) of certain more or less localized mappings, determinations, and decisions.

Related, of course, to determination, the issue of *agency* is central to a number of these essays. In Callon/Law's "Agency and the Hybrid *Collectif*," it is a matter of the taken-for-granted—but, as they point out, quite productively questioned—distinction between the human and the non-human. In Pickering's "Concepts and the Mangle of Practice: Constructing Quaternions," the issue is the relation between conceptual and material agency. The latter essay is a detailed analysis and reconstruction of the complex process by which a particular mathematical object was invented . . . or discovered . . . or constructed by the nineteenth-century mathematician William Rowan Hamilton. In view of the "but 2 + 2 still = 4" challenge mentioned above, it should be noted that Pickering's account requires neither the classical view of mathematical objects as prior and autonomous nor the idea—sometimes seen as the only alternative—that the technical decisions of scientists, including mathematicians, can be readily explained by political ideology or, in a simplistic sense, social interests. Mathematical objects are certainly understood here as socially, culturally, and discursively constituted, but through processes considerably subtler and more complexly mediated than generally thought, claimed, or caricatured. The conceptual difficulties and rhetorical risks involved in attempts to articulate such processes—and, more generally, to develop alternative accounts of agency and causality—are indicated both in the exchange between Pickering and Owen Flanagan and in the mutually frustrating confrontations discussed by Barbara Herrnstein Smith in her essay ("Microdynamics of Incommensurability: Philosophy of Science Meets Science Studies").

Issues of determination and agency surface as well in the contribution by Susan Oyama ("The Accidental Chordate: Contingency in Developmental Systems"), but here in connection with what are commonly represented as the genetic determinants of individual development. In questioning such representations in contemporary biology, Oyama plays out some implications of Darwin's (still radical enough) theory of evolution—which, it will be recalled, crucially foregoes appeal to either specific design or directionality. As she indicates, metaphors of control as well as teleology that recur in descriptions of the functioning of genes go some distance toward making the latter into updated, secular counterparts of that "ghost in the machine" (spirit, consciousness, intentionality, etc.) which, in theological/metaphysical philosophy of action, is thought to distinguish human agency from (mere) mechanical effectivity. (The agency of the individual subject, as traditionally conceived, is not Oyama's concern on this occasion, but the questions she raises here obviously affect the terms of that issue as well.) Revisionary ideas of human agency are also developed by Rotman, who redescribes mathematical activities as, in a rather more literal sense than usual, "thought experiments" — which, as such, involve a considerably expanded cast of subjective/"agential" characters, including "persons," "dreamers," and "imagoes," each with interesting relations to the others as well as highly differentiated roles in the relevant acts or scenes. Of course, from a quite traditional perspective it might be grumbled that agency, in this set of essays, is being intolerably fragmented or denied to everything or taken away from human beings uniquely and given to everything else — to pencils and assorted other lowly objects in Callon/Law's essay, to concepts and disciplinary discourses in Pickering's. It would be more accurate, however, to say that the idea of agency is evidently being, here (and in related work elsewhere), stretched, reconfigured, and transformed — or, in short, reconceptualized.

Allusions to *emergence,* emerging often, it appears, from biological models of temporal dynamics, recur in a number of these essays (and, it may be noticed, in this introduction as well). Their prominence seems to reflect the increasing need to address and describe, without implications of purposive agency or simple unilinear ("mechanical") causality, the ongoing effects of exceedingly complex interactions. Two points are worth stressing here.

First, postclassical theory has a distinctly Heraclitean flavor: these essays are dominated by images and models of flux — not simple mutability but complex and usually in(de)terminable dynamics, often involving significantly *reciprocal* interactions. Such images are evident not only in Oyama's piece, which is concerned with specifically biological systems and processes, but also in Pickering's "mangle" of mutually stabilizing interactive practices, in the interconnected lines and nodes of Callon/Law's "socio-technical networks," and in Smith's discussion of these and related theoretical developments. In all these cases, forces that are classically represented as distinct and opposed — for example, the genetic and the environmental or the natural and the social — are seen as reciprocally interactive and mutually constituting. This configuration of complex temporal dynamics and mutual constitution is also apparent in Roy Weintraub's account of the fortunes of a crucial theorem in mathematical economics ("Is 'Is a Precursor of' a Transitive Relation?"). As Weintraub indicates, what we take to be the events of intellectual history, including the history of science and mathematics, are not only continuously reinterpretable, but also, with respect to relations between causes and consequences, origins and effects, or "precursors" and "derivations," always potentially reversible.

Secondly, those always fundamentally unstable *constructs* and more or less radically contingent *constructings* or *interpretations*—which, here as elsewhere, replace the fixed entities-with-properties of traditional realism and the destined discovery-of-prior-facts of more traditional philosophy and history of science—seem to yield a universe where one can configure no reality *other* than virtual. Moreover, there appears to be no reason—the alleviation of cognitive dissonance aside—why there *should* be any other. That, at least, seems to have been the unscandalized attitude of Bohr, as discussed in Plotnitsky's contribution ("Complementarity, Idealization, and the Limits of Classical Conceptions of Reality"). If, as is often said, classical cosmology—as represented by post-Keplerian astronomy and Newton's mechanics—moved the human subject from a position of unique significance in the universe to that of a mere element in a machine which requires neither our participation nor even our existence to function, then postclassical physics can be said to restore the subject (so to speak) to significance, but in an epistemologically and otherwise disconcerting way. For the price of our apparently necessary participation in the constitution of the universe is the apparent dissolution of its *simple* existence—ourselves, of course, included. The price was, famously, too high for Einstein (among others) to be willing to pay. On the other hand, it could be thought that there are considerable—perhaps more than equal—conceptual and technical compensations to be gained from the exchange in question.

The concern with the agency of the non-human as well as the human in Callon/Law's essay, and with conceptual practices as well as material ones in Pickering's, reflects in both cases a commitment to *symmetry* that is an important feature of revisionist social studies of science and crucial to Malcolm Ashmore's essay ("Fraud by Numbers: Quantitative Rhetoric in the Piltdown Forgery Discovery"). As formulated especially by David Bloor and Barry Barnes, the so-called symmetry postulate of "the strong programme" in the sociology of science is not the (imagined) *epistemological* position that all truth-claims are equally (absolutely) valid, but the *methodological* position that, in accounting for the stabilization of knowledge, the sociologist (or historian) should not appeal to epistemically self-privileging or ("Whiggishly") present-privileging presumptions. Among such presumptions—the rejection of which would count as perverse skepticism from a classical perspective—are (a) that currently established knowledge reflects the uncovering of an always already determinate truth; (b) that the historical winners in scientific controversies are distinguished from the losers by the inherently better evidence they considered or the inherently better methods they used;

and (c) that genuine scientific knowledge can be distinguished from the products of bad science, pseudoscience, and fraudulent science on comparable evidential and/or methodological grounds *alone*. Thus Ashmore, in strict and scrupulous accordance with the symmetry postulate, subjects the analytic techniques by which the Piltdown forgery was exposed to the same detailed analysis that its self-declared "exposers" had used to undermine the claims made by the self-declared "discoverers" of certain contestably prehistoric bones. Like Weintraub and Pickering and in contrast to more traditional historians of science, Ashmore tells a pointedly non-Whiggish and non-teleological tale—and, in this case, also a pointedly non-edifying one.

From the perspective of a postclassical understanding of history, all the historical essays here "demonstrate" how scientific "facts" (such as those surrounding Piltdown Man) are produced by a continuous and quite complex play of forces, the traces of which are erased through the very process of the construction and appropriation of those facts *as such*—that is, without any quotation marks—by the relevant communities (scientists, engineers, historians, etc.). Thus, it could be said that Ashmore as *sociologist*, through his (deadpan) rigorously scientific analysis, demonstrates how the standards of individual scientists are sometimes less than rigorously scientific. But, applying the symmetry principle reflexively, Ashmore as *historian* could no more claim to have demonstrated (or "recovered") the *historical* facts of the Piltdown case than any of the scientists in question could claim to have demonstrated (or "discovered") the *scientific* facts. The point to be stressed is not that demonstrations, including rigorous ones, are impossible (a frequent misunderstanding), but that classical conceptions of what constitutes a demonstration—and, indeed, what constitutes rigor—cannot be taken for granted. What this means most significantly here is that no demonstration, whether unselfconsciously asymmetrical or rigorously symmetrical, can escape the multifarious play of forces, although these forces may, in each case, be played out quite differently and also constitute themselves quite differently to begin with.

Symmetries—exact and broken—are at the heart of the conceptual patterns, puzzles, and paradoxes that link the projects (or perhaps obsessions) of mathematical logic with the endless fascination which the joint instabilities of number and word, truth and lie, have held for a number of writers, ancient to postmodern. In addition to Shakespeare and Beckett, whose imaginative fertility and conceptual precision (and radicality) in this respect Smyth delineates in his essay ("A Glance at SunSet: Numerical Fundaments in Frege, Wittgenstein, Shakespeare, Beckett"), one could also cite James Joyce, from whose *Finnegans Wake* the term "quark" was obtained, courtesy of the physicist Murray Gell-Mann. Like Beckett, Joyce was absorbed by the paradoxical strangeness of twentieth-century logic and science — just as Wittgenstein, Gell-Mann, and other key figures of modern philosophy and science were absorbed in turn by equally strange and paradoxical aspects of twentieth-century literature and art. Notice of such intellectual relations and individual ranges ("influences," "borrowings," "polymaths," etc.) is not unusual in literary and cultural studies. It is commonly confined, however, to particular cases and serves, often enough, to reinforce the very oppositional pairings — that is, art and philosophy, philosophy and science, science and literature, literature and logic, and so forth — that such cases might otherwise be thought to unsettle. In this respect and others, Smyth's essay, which considers the manifold of decidable and undecidable combinations, parallel situations, and trajectories among the constituents of all these classic pairs, demonstrates the impossibility of fixing any such sets of relations and offers a more aptly postclassical exploration.

Joining many of the recurrent themes and conceptual structures discussed above is the idea of *undecidability*. The latter term is borrowed from post-Gödelian mathematical logic where, in its most general and radical conception, it designates propositions that can be formulated within a given system but whose truth or falsity can never be established by means available in the system itself. Gödel proved that any formal system of axioms and rules, if it is large enough to contain arithmetic and, importantly, if it is free from contradiction, must contain some statements that are neither provable nor disprovable within the system. The truth of such statements is thus *undecidable* by standard procedures. (In fact, Gödel proved that the statement concerning the consistency of such a system is itself undecidable. It follows that such a system, although it can never be formally, mathematically proved to be consistent, could eventually be discovered to contain a contradiction; thus, there is no guarantee that the proposition "2 + 2 = 5" could not one day be derived from a system of axioms containing arithmetic.)

Conceptual and metaphorical analogies with Gödel's undecidables have been effectively deployed by a number of postclassical theorists (notably, Derrida), and the idea resurfaces more or less explicitly in several contributions here. Callon and Law, for example, write as follows:

> Debates about the status of agency are metaphysical. Are "humans" "like" "non-humans," or not? This is undecidable. Or perhaps it can

sometimes be decided, but only locally. So we can't in general *prove*
that "humans" are like "non-humans." Or that some "non-humans"
are agents. . . . All we can do is make stories which suggest that
if you don't make such assumptions [i.e., that non-human agency
is a "contradiction in terms"], then revealing things may happen,
theoretically and empirically.

Here one might say, by analogy with Gödel, that the truth or falsity of
metaphysical propositions about agency cannot be ascertained *in general*. Locally, however (i.e., in a particular situation), assuming the truth
or falsity of a given proposition—with or without granting the corresponding metaphysical assumption—may have significant theoretical
and practical consequences. This distinctively postclassical "logic" is applied by Derrida to a number of specific instances, such as the classic
opposition between literature and philosophy (in his reading of Mal-
larmé in *Dissemination*), and also very generally to what he refers to as
"all the pairs of opposites on which philosophy is constructed," which
would include culture and nature, mind and matter, concepts and intuitions, and writing and speech. In accord with the "logic" just indicated,
both the difference and the hierarchy between the constituents of each
pair would be seen as generally, but not unconditionally, "undecidable."
In the logic of most philosophical systems, however, both the difference and the hierarchy would be seen as absolute and decidable—for
example, nature over culture, mind over matter, or intuitions over concepts—although, significantly enough, in different philosophical systems (and at different points in the same system) these hierarchies are
produced in reverse, which in part enables Derrida to conclude that any
such decidability can only be local or provisional.

An opposition of particular interest here is the one between philosophical and mathematical knowledge, considered briefly above. Are
they fundamentally different? There are, as it happens, many classical
answers, some of them mutually exclusive or conflicting, each claiming
to decide the question once and for all, with many ensuing definitions
of mathematics and philosophy and also the corresponding hierarchical
relations between them (especially with respect to their proximity to
some posited ultimate truth or knowledge, such as God's or that of a
computer). Postclassical theory would see the question as, for many reasons—historical, theoretical, cultural, and other—undecidable *in general*, that is, unconditionally or once and for all, although decidable
(and evidently decided) in specific situations. The same type of question can be asked of the relation between mathematical knowledge and

literature, or between science and art more generally, and answered in comparable ways. Smyth's essay demonstrates that the "logic" of numbers may itself follow many logics (some of which are illogical in classical terms); that mathematics can sometimes become literary and, conversely, that literature can and sometimes must be mathematical; and that all these relations can be decided differently under different circumstances. In the domains of either literature or mathematics, theater or life, the relation between mathematics and literature or (as Shakespeare famously told us) between theater and life cannot be established or decided once and for all. This is not to say that all differences or decisions between them are suspended as a result, but that any one of them (literature, life, art, mathematics, etc.) can be put into play in relation to the others in ways very different from those given by classical theory.

It appears, then, that all general metaphysical propositions, including those concerning the philosophical foundations of mathematics — or of physics or other sciences, or of literature or philosophy itself, or of the relationships between and among all these — are undecidable at best: that is, as Nietzsche was fond of pointing out, they can never be claimed true as general propositions, not even by any classical criterion of truth (a concept which can itself be defined, so to speak, only by means of undecidable propositions). This, again, does not eliminate locally significant differences between such propositions or the particular metaphysical assumptions correlative to them: such differences are often (though not always) decidable and sometimes also decisive. It remains the case, however, that some among such metaphysical assumptions may need to be abandoned altogether by a given postclassical theory. Nietzsche certainly abandoned a great many of them, but still far from all of them.

One implication of this general situation for the various projects of postclassical theory is the possibility, and sometimes the necessity, of utilizing mutually exclusive concepts (or concepts that appear to be "contradictions in terms") within the same framework, and without a classical (for example, Hegelian) synthesis. Niels Bohr's interpretation of quantum mechanics as what he called *complementarity*, discussed in some detail here by Plotnitsky, is precisely such an instance of necessity. Complementarity is not itself undecidability as conceived on the Gödelian model. (Nor, for that matter, is *indeterminacy* as represented by Heisenberg's uncertainty relations.) One can say, however, that complementarity is made possible by undecidability insofar as the latter concept applies to general metaphysical propositions and assumptions, as opposed to strictly mathematical ones, as in Gödel. Complementarity configurations of the former type—that is, arising from the undecidability of (or the impossibility of claiming general truth for) metaphysical assumptions—may be detected in several essays in this collection. For example, Callon/Law's "hybrid *collectif*" could be seen as an undecidable/complementary configuration of human and non-human participants in laboratory life, replacing classical definitions and related oppositional distinctions of agency. Analogous configurations are both implicit and explicit in Oyama's account of the interplay of chance and necessity, contingency and regularity, and predictability and unpredictability, both in evolutionary theory and in the development of individual organisms; such configurations are also evident in her account of the interplay among the different constituents of these classical oppositional pairs, which are not always equivalent to each other. As observed earlier, ideas of emergence and models of complex and often reciprocal processes dominate several essays in this collection, including the two just mentioned. It is worth adding that, in this respect, they differ interestingly and significantly from the more static configuration of the mathematical model of undecidability per se. But that confirms the point being made here, for in this respect these essays demonstrate the possible connections between, on the one side, undecidability or complementarity and, on the other, process, emergence, becoming, or history.

A final, reflexive point should be stressed here, namely, that the post-classical logic of undecidability may be applied to the very opposition between classical and postclassical. For this opposition, too, cannot be established once and for all, either theoretically or historically; nor can any hierarchy be established unconditionally between its constituents. Across the spectrum of the history or, speaking more postclassically, *histories* of modern mathematics, science, and theory—from Galois to Hamilton to Einstein to Bohr to Gödel; from classical algebra to quaternions to non-Euclidean geometry to Einsteinian relativity to quantum theory to chaos theory; from Adam Smith to the Arrow theorem and the complex mathematical models of modern economics; from Darwin to the neo-Darwinian synthesis to contemporary genetics to developmental systems theory; from the work of individual mathematicians and physicists, such as those mentioned above, to modern experimental science to postmodern big science and laboratory life; from Pascal's first calculating machine to modern computers to virtual reality, artificial intelligence, and artificial life; and from Hegel to Marx to Nietzsche to Wittgenstein to Feyerabend to Foucault to Derrida to the authors of the essays collected here: in all these fields and in the thinking of all the figures involved, including our own interpretations of these events and histories, there is an immensely complex and sometimes undecidable interplay between that which is classical and that which is postclassical. For the moment, however, one might want to stress the local decidability and decisiveness of certain differences (often radical, though of course never absolute) between them, for example, the very logic of undecidability just delineated. This logic, which does not appear to be found in any classical theory, at least not in this form, appears to be decidably and decisively *postclassical*.

Recalling the dissatisfaction with classical conceptions of knowledge, language, truth, proof, reality, and representation (etc.) noted at the beginning, one could say that at stake in the *decision* here is a transformation of our understanding of the operations and significance of mathematics, science, and theory, classical and postclassical alike. In that respect, one might see the value of this collection as its demonstration that interest in the reconceptualizations that define postclassical theory from the era of Bohr and Wittgenstein to our own has moved from the question of intellectual legitimacy or philosophical scandal to the question of what, specifically, can be done with them, that is, to institutional practices. Once a theory, conceptualization, or reconceptualization begins to be played out in interesting, productive, and connectible ways in a number of domains of knowledge, such as biology, the history and sociology of science, literary theory, mathematics, and philosophy, then the question of its legitimacy has already been answered and so, too, one might say, has the question of its validity.

Acknowledgments. This collection of essays, first published as volume 94, number 2 of the *South Atlantic Quarterly* (without the final essay and index), originated in a colloquium on Mathematics and Postclassical Theory that took place at Duke University in fall 1993. The degree of interdisciplinary collegiality exhibited on that occasion — including the emergence during the course of the proceedings of a sort of intellectual "pidgin" language, initially awkward and even comical but ultimately quite effective — was remarked by participants and audience alike. The success of the colloquium in those and other respects reflected the energies, enthusiasms, and adventuresomeness of a number of people, and the existence of this volume reflects their willingness to extend even further their commitment to the general enterprise it represented. In addition to the participants whose essays appear in this volume — and the subsequent contributors: Malcolm Ashmore, Michel Callon, John Law, and Susan Oyama — we would like to acknowledge in particular

Robert Bryant and David Morrison, both members of Duke University's Department of Mathematics, whose contributions to the colloquium's proceedings were wonderful testimony to the intellectual value of such attempted exchanges across traditional disciplinary borders; and Fredric Jameson, editor of *SAQ*, whose continuing interest, support, and encouragement were in large measure responsible for the colloquium's afterlife as represented by this volume.

Thinking Dia-Grams: Mathematics,

Writing, and Virtual Reality

Brian Rotman

I N the epilogue to his essay on the development of writing systems, Roy Harris declares:

> It says a great deal about Western culture that the question of the origin of writing could be posed clearly for the first time only after the traditional dogmas about the relationship between speech and writing had been subjected both to the brash counterpropaganda of a McLuhan and to the inquisitorial scepticism of a Derrida. But it says even more that the question could not be posed clearly until writing itself had dwindled to microchip dimensions. Only with this . . . did it become obvious that the origin of writing must be linked to the future of writing in ways which bypass speech altogether.[1]

Harris's intent is programmatic. The passage continues with the injunction not to "re-plough McLuhan's field, or Derrida's either," but sow them, so as to produce eventually a "history of writing *as writing*."

Preeminent among dogmas that block such a history is alphabeticism: the insistence that we interpret all writing—understood for the moment as any systematized graphic activity that creates sites of interpretation and facilitates communication and sense making—along the lines of alphabetic writing, as if it were the inscription of prior speech ("prior" in an ontogenetic sense as well as the more immediate sense of speech first uttered and then written down and recorded). Harris's own writings in linguistics as well as Derrida's program of deconstruction, McLuhan's efforts to dramatize the cultural imprisonments of typography, and Walter Ong's long-standing theorization of the orality/writing disjunction in relation to consciousness, among others, have all demonstrated the distorting and reductive effects of the subordination of graphics to phonetics and have made it their business to move beyond this dogma. Whether, as Harris intimates, writing will one day find a speechless characterization of itself is impossible to know, but these displacements of the alphabet's hegemony have already resulted in an open-ended and more complex articulation of the writing/speech couple, especially in relation to human consciousness, than was thinkable before the microchip.

A written symbol long recognized as operating nonalphabetically — even by those deeply and quite unconsciously committed to alphabeticism — is that of number, the familiar and simple other half, as it were, of the alphanumeric keyboard. But, despite this recognition, there has been no sustained attention to mathematical writing even remotely matching the enormous outpouring of analysis, philosophizing, and deconstructive opening up of what those in the humanities have come simply to call "texts."

Why, one might ask, should this be so? Why should the sign system long acknowledged as the paradigm of abstract rational thought and the without-which-nothing of Western technoscience have been so unexamined, let alone analyzed, theorized, or deconstructed, as a mode of writing? One answer might be a second-order or reflexive version of Harris's point about the microchip dwindling of writing, since the very emergence of the microchip is inseparable from the action and character of mathematical writing. Not only would the entire computer revolution have been impossible without mathematics as the enabling conceptual technology (the same could be said in one way or another of all technoscience), but, more crucially, the computer's mathematical lineage and intended application as a calculating/reasoning machine hinges on its autological relation to mathematical practice. Given this autology, mathematics would presumably be the last to reveal itself and declare its origins in writing. (I shall return to this later.)

A quite different and more immediate answer stems from the difficulties put in the way of any proper examination of mathematical writing by the traditional characterizations of mathematics — Platonic realism or various intuitionisms — and by the moves they have legitimated within the mathematical community. Platonism is the contemporary orthodoxy. In its standard version it holds that mathematical objects are mentally apprehensible and yet owe nothing to human culture; they exist, are real, objective, and "out there," yet are without material, empirical, embodied, or sensory dimension. Besides making an enigma out of mathematics' usefulness, this has the consequence of denying or marginalizing to the point of travesty the ways in which mathematical signs are the means by which communication, significance, and semiosis are *brought about.* In other words, the constitutive nature of mathematical writing is invisibilized, mathematical language in general being seen as a neutral and inert medium for describing a given prior reality — such as that of number — to which it is essentially and irremediably posterior.

With intuitionist viewpoints such as those of Brouwer and Husserl, the source of the difficulty is not understood in terms of some *external* metaphysical reality, but rather as the nature of our supposed internal intuition of mathematical objects. In Brouwer's case this is settled at the outset: numbers are nothing other than ideal objects formed within the inner Kantian intuition of time that is the condition for the possibility of our cognition, which leads Brouwer into the quasi-solipsistic position that mathematics is an essentially "languageless activity." With Husserl, whose account of intuition, language, and ideality is a great deal more elaborated than Brouwer's, the end result is nonetheless a complete blindness to the creative and generative role played by mathematical writing. Thus, in "The Origin of Geometry," the central puzzle on which Husserl meditates is "How does geometrical ideality . . . proceed from its primary intrapersonal origin, where it is a structure within the conscious space of the first inventor's soul, to its ideal objectivity?"[2] It must be said that Husserl doesn't, in this essay or anywhere else, settle his question. And one suspects that it is incapable of solution. Rather, it is the premise itself that has to be denied: that is, it is the coherence of the idea of primal (semiotically unmediated) intuition lodged originally in any individual consciousness that has to be rejected. On the contrary, does not all mathematical intuition — geometrical or otherwise — come into being in relation to mathematical signs, making it both external/intersubjective and internal/private from the start? But to pursue such a line one has to credit writing with more than a capacity to, as Husserl has it, "document," "record," and "awaken" a prior and necessarily pre-linguistic mathematical meaning. And this is precisely what his whole understanding of language and his picture of the "objectivity" of the ideal prevents him from doing. One consequence of what we might call the documentist view of mathematical writing, whether Husserl's or the standard Platonic version, is that the intricate interplay of imagining and symbolizing, familiar on an everyday basis to mathematicians within their practice, goes unseen.

Nowhere is the documentist understanding of mathematical language more profoundly embraced than in the foundations of mathematics, specifically, in the Platonist program of rigor instituted by twentieth-century mathematical logic. Here the aim has been to show how all of mathematics can be construed as being about sets and, further, can be translated into axiomatic set theory. The procedure is twofold. First, vernacular mathematical usage is made informally rigorous by having all of its terms translated into the language of sets. Second, these informal translations are completely formalized, that is, further translated into an axiomatic system consisting of a Fregean first-order logic supplemented with the extralogical symbol for set membership.

To illustrate, let the vernacular item be the theorem of Euclidean geometry which asserts that, given any triangle in the plane, one can draw a unique inscribed circle (see diagram at right). The first translation removes all reference to agency, modality, and physical activity, signaled here in the expression "one can draw." In their place are constructs written in the timeless and agentless language of sets. Thus, first the plane is identified with the set of all ordered pairs (x, y) of real numbers and a line and circle are translated into certain unambiguously determined subsets of these ordered pairs through their standard Cartesian equations; then "triangle" is rendered as a triple of nonparallel lines and "inscribed" is given in terms of a "tangent," which is explicated as a line intersecting that which it "touches" in exactly one point. The second translation converts the asserted relationship between these abstracted but still visualizable sets into the de-physicalized and de-contextualized logico-syntactical form known as the first-order language of set theory. This will employ no linguistic resources whatsoever other than variables ranging over real numbers, the membership relation between sets, the signs for an ordered pair and for equality, the quantifiers "for all x" and "there exists x," and the sentential connectives "or," "and," "not," and so on.

Once such a double translation of mathematics is effected, metamathematics becomes possible, since one can arrive at results about the whole of vernacular mathematics by proving theorems about the formal (i.e., mathematized) axiomatic system. The outcome has been an influential and rich corpus of metamathematical theorems (associated with Skolem, Gödel, Turing, and Cohen, among many, many others). Philosophically, however, the original purpose of the whole foundational enterprise was to illuminate the nature of mathematics by explaining the emergence of paradox, clarifying the horizons of mathematical reasoning, and revealing the status of mathematical objects. In relation to these aims the set theoretization of mathematics and the technical results of metamathematics are unimpressive: not only have they resulted in what is generally acknowledged to be a barren and uninformative philosophy of mathematics, but (not independently) they have failed to

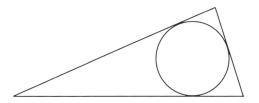

shed any light whatsoever on mathematics as a signifying practice. We need, then, to explain the reason for this impoverishment.

Elsewhere, I've spelled out a semiotic account of mathematics, particularly the interplay of writing and thinking, by developing a model of mathematical activity—what it means to make the signs and think the thoughts of mathematics—intended to be recognizable to its practitioners.[3] The model is based on the semiotics of Charles Sanders Peirce, which grew out of his program of pragmaticism, the general insistence that "the meaning and essence of every conception lies in the application that is to be made of it."[4] He understood signs accordingly in terms of the uses we make of them, a sign being something always involving another—interpreting—sign in a process that leads back eventually to its application in our lives by way of a modification of our habitual responses to the world. We acquire new habits in order to minimize the unexpected and the unforeseen, to defend ourselves "from the angles of hard fact" that reality and brute experience are so adept at providing. Thought, at least in its empirically useful form, thus becomes a kind of mental experimentation, the perpetual imagining and rehearsal of unforeseen circumstances and situations of possible danger. Peirce's notion of habit and his definition of a sign are rich, productive, and capable of much interpretation. They have also been much criticized; his insistence on portraying all instances of reasoning as so many different forms of disaster avoidance is obviously unacceptably limiting. In this connection, Samuel Weber has made the suggestion that Peirce's "attempt to construe thinking and meaning in terms of 'conditional possibility,' and thus to extend controlled laboratory experimentation into a model of thinking in general," should be seen as an articulation of a "phobic mode of behavior," where the fear is that of ambiguity in the form of cognitive oscillation or irresolution, blurring or shifting of boundaries, imprecision, or any departure from the clarity and determinateness of either/or logic.[5]

Now, it is precisely the elimination of these phobia-inducing features that reigns supreme within mathematics. Unashamedly so: mathematicians would deny that their fears were pathologies, but would, on the contrary, see them as producing what is cognitively and aesthetically attractive about mathematical practice as well as being the source of its utility and transcultural stability. This being so, a model of mathematics utilizing the semiotic insights of Peirce—himself a mathematician—might indeed deliver something recognizable to those who practice mathematics. The procedure is not, however, without risks. There is evidently a self-confirmatory loop at work in the idea of using such a theory to illuminate mathematics, in applying a phobically derived apparatus, as it were, to explicate an unrepentant instance of itself. In relation to this, it is worth remarking that Peirce's contemporary, Ernst Mach, argued for the importance of thought-experimental reasoning to science from a viewpoint quite different from Peirce's semiotics, namely, that of the physicist. Indeed, thought experiments have been central to scientific persuasion and explication from Galileo to the present, figuring decisively in this century, for example, in the original presentation of relativity theory as well as in the Einstein-Bohr debate about the nature of quantum physics. They have, however, only recently been given the sort of sustained attention they deserve. Doubtless, part of the explanation for this comparative neglect of experimental reasoning lies in the systematizing approach to the philosophy of science that has foregrounded questions of rigor (certitude, epistemological hygiene, formal foundations, exact knowledge, and so on) at the expense of everything else, and in particular at the expense of any account of the all-important persuasive, rhetorical, and semiotic content of scientific practice.

In any event, the model I propose theorizes mathematical reasoning and persuasion in terms of the performing of thought experiments or waking dreams: one does mathematics by propelling an imago—an idealized version of oneself that Peirce called the "skeleton self"—around an imagined landscape of signs. This model depicts mathematics, by which I mean here the everyday doing of mathematics, as a certain kind of traffic with symbols, *a written discourse* in other words, as follows: All mathematical activity takes place in relation to three interlinked arenas—Code, MetaCode, Virtual Code. These represent three complementary facets of mathematical discourse; each is associated with a semiotically defined abstraction, or linguistic actor—Subject, Person, Agent, respectively—that "speaks," or uses, it. The following diagram summarizes these actors and the arenas in relation to which they operate as an interlinked triad. The Code embraces the total of

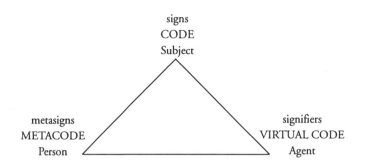

all rigorous sign practices — defining, proving, notating, and manipulating symbols — sanctioned by the mathematical community. The Code's user, the one-who-speaks it, is the mathematical *Subject*. The Subject is the agency who reads/writes mathematical texts and has access to all and only those linguistic means allowed by the Code. The MetaCode is the entire matrix of unrigorous mathematical procedures normally thought of as preparatory and epiphenomenal to the real — proper, rigorous — business of doing mathematics. Included in the MetaCode's resources would be the stories, motives, pictures, diagrams, and other so-called heuristics which introduce, explain, naturalize, legitimate, clarify, and furnish the point of the notations and logical moves that control the operations of the Code. The one-who-speaks the MetaCode, the *Person*, is envisaged as being immersed in natural language, with access to its metasigns and constituted thereby as a self-conscious subjectivity in history and culture. Lastly, the Virtual Code is understood as the domain of all legitimately imaginable operations, that is, as signifying possibilities available to an idealization of the Subject. This idealization, the one-who-executes these activities, the *Agent*, is envisaged as a surrogate or proxy of the Subject, imagined into being precisely in order to act on the purely formal, mechanically specifiable correlates — signifiers — of what for the Subject is meaningful via signs. In unison, these three agencies make up what we ordinarily call "the mathematician."

Mathematical reasoning is thus an irreducibly tripartite activity in which the Person (Dreamer awake) observes the Subject (Dreamer) imagining a proxy — the Agent (Imago) — of him/herself, and, on the basis of the likeness between Subject and Agent, comes to be persuaded that what the Agent experiences is what the Subject *would* experience were he or she to carry out the unidealized versions of the activities in question. We might observe in passing that the three-way process at work here is the logico-mathematical correlate of a more general and originating triangularity inherent to the usual divisions invoked to articulate self-consciousness: the self-as-object instantiated here by the Agent, the self-as-subject by the Subject, and the sociocultural other, through which any such circuit of selves passes, by the Person.

Two features of this way of understanding mathematical activity are relevant here: First, mathematical assertions are to be seen, as Peirce insisted, as foretellings, predictions made by the Person about the Subject's future engagement with signs, with the result that the process of persuasion is impossible to comprehend if the role of the Person as observer of the Subject/Agent relation is omitted. Second, mathematical thinking and writing are folded into each other and are inseparable not only in an obvious practical sense, but also theoretically, in relation to the cognitive possibilities that are mathematically available. This is because the Agent's activities exist and make narrative logical sense (for the Subject) only through the Subject's manipulation of signs in the Code.

The second feature of this model, the thinking/writing nexus, will occupy us below. On the first, however, observe that there is an evident relation between the triad of Code/MetaCode/Virtual Code here and the three levels — rigorous/vernacular/formal — of the Platonistic reduction illustrated above. Indeed, in terms of external attributes, the difference drawn by the mathematical community between unrigorous/vernacular and rigorous/set-theoretical mathematics seems to map onto that between the MetaCode and the Code. This is indeed the case, but the *status* of this difference is here inverted and displaced. On the present account, belief in the validity of reasoning, or acquiescence in proof, takes place only when the Person is persuaded, a process that hinges on a judgment — available only to the Person — that the likeness between the Subject and the Agent justifies replacing the former with the latter. In terms of our example, the proof of the theorem in question lies in the relationship among the Person, who can draw a triangle and see it as a drawn triangle; the Subject, who can replace this triangle with a set-theoretical description; and the Agent, who can act upon an imagined version of this triangle. By removing all reference to agency, the Platonistic account renders this triple relation invisible. Put differently, in the absence of the Person's role, no explication of *conviction* — without which proofs are not proofs — can be given. Instead, all one can say about a supposed proof is that its steps, as performed by the "mathe-

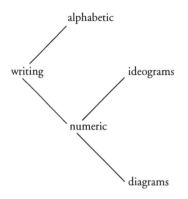

matician," are logically correct, a truncated and wholly unilluminating description of mathematical reasoning found and uncritically repeated in most contemporary mathematical and philosophical accounts. But once the Person is acknowledged as vital to the mathematical activities of making and proving assertions, it becomes impossible to see the MetaCode as a supplement to the Code, as a domain of mere psychological/motivational affect, to be jettisoned as soon as the real, proper, rigorous mathematics of the Code has been formulated.

If we grant this, we are faced with a crucial difference operating within what are normally and uncritically called mathematical "symbols," a difference whose status is not only misperceived within the contemporary Platonistic program of rigor, but, beyond this, is treated within that program to a reductive alphabeticization. The situation is encapsulated in the above diagram. The first split, the division of writing into the alpha and the numeric, is simply the standard recognition of the nonalphabetic character of numeric, that is, mathematical, writing. The failure to make this distinction, or rather making it but subsuming all writing under the rubric of the alphabet, is merely an instance of what we earlier called the alphabetic dogma. The point of the latter diagram is to indicate the replication of this split, via a transposed version of itself, within the contemporary Platonistic understanding of mathematics. Thus, on one hand, there are ideograms, such as '+', '×', '1', '2', '3', '=', '>', '. . .', 'sinz', 'logz', and so on, whose introduction and interaction are controlled by rigorously specified rules and syntactic conventions. On the other hand, there are diagrams, visually presented semiotic devices, such as the familiar lines, axes, points, circles, and triangles, as well as all manner of figures, markers, graphs, charts, commuting circuits, and iconically intended shapes. On the orthodox view, the difference is akin to that between the rigorously literal, clear, and unambiguous ideograms and the metaphorically unrigorous diagrams. The transposition in question is evident once one puts this ranking of the literal over the metaphorical into play: as soon, that is, as one accepts the idea that diagrams, however useful and apparently essential for the actual doing of mathematics, are nonetheless merely figurative and eliminable and that mathematics, in its proper rigorous formulation, has no need of them. Within the Platonist program, this alphabetic prejudice is given a literal manifestation: linear strings of symbols in the form of normalized sequences of variables and logical connectives drawn from a short, preset list determine the resting place for mathematical language in its purest, most rigorously grounded form.

There is a philosophical connection between this transposed alphabeticism and classical ontology. The alphabetic dogma rides on and promotes an essential secondarity. In its original form, this meant the priority of speech to writing, that is, the insistence that writing is the transcription of an always preceding speech (and, taking the dogma further back, that speech is the expression of a prior thought, which in turn is the mirror of a prior realm, . . .). Current Platonistic interpretations of mathematical signs replay this secondarity by insisting that signs are always signs *of* or *about* some preexisting domain of objects. Thus the time-honored distinction between numerals and numbers rests on just such an insistence that numerals are mere notations — names — subsequent and posterior to numbers which exist prior to and independent of them. According to this understanding of signs, it's easy to concede that numerals are historically invented, changeable, contingent, and very much a human product, while maintaining a total and well-defended refusal to allow any of these characteristics to apply to numbers. And what goes for numbers goes for all mathematical objects. In short, contemporary Platonism's interpretation of rigor and its ontology — despite appearances, manner of presentation, and declared motive — go hand in glove: both relying on and indeed constituted by the twin poles of an assumed and never-questioned secondarity.

Similar considerations are at work within Husserl's phenomenological project and its problematic of geometrical origins. Only there, the pre-semiotic — that which is supposed to precede all mathematical language — is not a domain of external Platonic objects subsequently described by mathematical signs, but a field of intuition. The prior scene is one of "primal intrapersonal intuition" that is somehow — and this is

Husserl's insoluble problem — "awakened" and "reactivated" by mathematical writing in order to become available to all men at all times as an objective, unchanging ideality. The problem evaporates as a mere misperception, however, if mathematical writing is seen not as secondary and posterior to a privately engendered intuition, but as constitutive of and folded into the mathematical meaning attached to such a notion. What was private and intrapersonal is revealed as already intersubjective and public.

Nowhere is this more so than in the case of diagrams. The intrinsic difficulty diagrams pose to Platonistic rigor — their essential difference from abstractly conceived sets and the consequent need to replace them with ideogrammatic representations — results in their elimination in the passage from vernacular to formalized mathematics. And indeed, set-theoretical rewritings of mathematics (notably Bourbaki's but in truth almost all contemporary rigorous presentations of the subject) avoid diagrams like the plague. And Husserl, for all his critique of Platonistic metaphysics, is no different: not only are diagrams absent from his discussion of the nature of geometry, a strange omission in itself, but they don't even figure as an important item to thematize.

Why should this be so? Why, from divergent perspectives and aims, should both Platonism and Husserlian phenomenology avoid all figures, pictures, and visual inscriptions in this way? One answer is that diagrams — whether actual figures drawn on the page or their imagined versions — are the work of the body; they are created and maintained as entities and attain significance only in relation to human visual-kinetic presence, only in relation to our experience of the culturally inflected world. As such, they not only introduce the historical contingency inherent to all cultural activity, but, more to the present point, they call attention to the materiality of all signs and of the corporeality of those who manipulate them in a way that ideograms — which appear to denote purely "mental" entities — do not. And neither Platonism's belief in timeless transcendental truth nor phenomenology's search for ideal objectivity, both irremediably mentalistic, can survive such an incursion of physicality. In other words, diagrams are inseparable from perception: only on the basis of our encounters with actual figures can we have any cognitive or mathematical relation to their idealized forms. The triangle-as-geometrical-object that Husserl would ignore, or Platonists eliminate from mathematics proper, is not only what makes it possible to think that there could be a purely abstract or formal or mental triangle, but is also an always available point of return for geometrical abstraction that ensures its never being abstracted out of the frame of mathematical discourse. For Merleau-Ponty the necessity of this encounter is the essence of a diagram:

> I believe that the triangle has always had, and always will have, angles the sum of which equals two right angles . . . because I have had the experience of a real triangle, and because, as a physical thing, it necessarily has within itself everything it has ever been able, or ever will be able, to display. . . . What I call the essence of the triangle is nothing but this presumption of a completed synthesis, in terms of which we have defined the thing.[6]

In fact, the triangle and its generalizations constitute geometry as means as well as object of investigation: geometry is a mode of imagining with and about diagrams.

And indeed, one finds recent commentaries on mathematics indicating a recognition of the diagrammatic as opposed to the purely formal nature of mathematical intuition. Thus Philip Davis urges a reinterpretation of "theorem" which would include visual aspects of mathematical thought occluded by the prevailing set-theoretical rigor, and he cites V. I. Arnold's repudiation of the scorn with which the Bourbaki collective proclaims that, unlike earlier mathematical works, *its* thousands of pages contain not a single diagram.[7]

Let us return to our starting point, the question of writing, or, as Harris puts it, "writing *as writing*." One of the consequences of my model is to open up mathematical writing in a direction familiar to those in the humanities. As soon as it becomes clear that diagrams (and indeed all the semiotic devices and sources of intuition mobilized by the MetaCode) can no longer be thought of as the unrigorous penumbra of proper — Coded — mathematics, as so many ladders to be kicked away once the ascent to pure, perceptionless Platonic form has been realized, then all manner of possibilities can emerge. All we need to do to facilitate them is to accept a revaluation of basic terms. Thus, there is nothing intrinsically wrong with or undesirable about "rigor" in mathematics. Far from it: without rigor, mathematics would vanish; the question is how one interprets its scope and purpose. In the present account, rigor is not an externally enforced program of foundational hygiene, but an intrinsic and inescapable demand proceeding from writing: it lies within the rules, conventions, dictates, protocols, and such that control mathematical imagination and transform mathematical intuition into an intersubjective writing/thinking practice. It is in this sense that, for example, Gregory Bateson's tag line that mathematics is a world of "rigorous fantasy" should be read. Likewise, one can grant that the MetaCode/Code division is akin to the metaphor/literal opposition, but refuse the pejorative sense that set-theoretical rigor has assigned to the term "metaphor." Of course, there is a price to pay. Discussions of tropes in the humanities have revealed that no simple or final solution to the "problem" of metaphor is possible; there is an always uneliminable reflexivity, since it proves impossible, in fact and in principle, to find a trope-free metalanguage in which to discuss tropes and so to explain metaphor in terms of something nonmetaphorical. For mathematics the price — if such it be — is the end of the foundational ambition, the desire to ground mathematics, once and for all, in something fixed, totally certain, timeless, and prelinguistic. Mathematics is not a building — an edifice of knowledge whose truth and certainty is guaranteed by an ultimate and unshakable support — but a process: an ongoing, open-ended, highly controlled, and specific form of written intersubjectivity.

What, then, are we to make of mathematical diagrams, of their status as writing? How are they to be characterized vis-à-vis mathematical ideograms, on the one hand, and the words of nonmathematical texts, on the other? It would be tempting to invoke Peirce's celebrated trichotomy of signs — symbol, icon, and index — at this point. One could ignore indexicals and regard ideograms as symbols (signs resting on an arbitrary relation between signifier and signified) and diagrams as icons (signs resting on a motivated connection between the two). Although there is truth in such a division, it is a misleading simplification: the ideogram/diagram split maps only with great artificiality onto these two terms of Peirce's triad. In addition, there is a terminological difficulty: Peirce restricted the term "diagram" to one of three kinds of icon (the others being "image" and "metaphor"), which makes his usage too narrow for what we here, and mathematicians generally, call diagrams. The artificiality arises from the fact that the ideogrammatic cannot be separated from the iconic nor the diagrammatic from the indexical. Thus, not only are ideograms often enmeshed in iconic sign use at the level of algebraic schemata, but, more crucially, diagrams, though iconic, are also, less obviously, indexical to varying degrees. Indeed, the very fact of their being physically experienced shapes, of their having an operative meaning inseparable from an embodied and therefore situated gesture, will ensure that this is the case.

But this is a very generic source of indexicality, and some diagrams exhibit much stronger instances. Thus, consider the diagram, fundamental to post-Renaissance mathematics, of a coordinate axis which

consists of an extended, directed line and an origin denoted by zero. Let's ignore the important but diffuse indexicality brought into play by the idea of directed extension and focus on that of the origin. Clearly, the function of the ideogram 'o' in this diagram is to establish an arbitrarily chosen but fixed and distinguished "here" within the undifferentiated linear continuum. The ideogram marks a "this" with respect to which all positions on the line can be oriented; such is what it means for a sign to function as an origin of coordinates. Indexicality, interpreted in the usual way as a coupling of utterance and physical circumstance, and recognized as present in the use of shifters like "this," "here," "now," and so on, within ordinary language, is thus unambiguously present in our diagram. It does not, however, declare itself as such: its presence is the result of a choice and a determination made in the MetaCode, that is, outside the various uses of the diagram sanctioned within mathematics proper. It is, in other words, the written evidence or trace of an originating act by the Person. Thus, zero, when symbolized by 'o', is an ideogrammatic sign for mathematical 'nothing' at the same time that it performs a quasi-indexical function within the diagram of a coordinate axis.[8]

It would follow from considerations like these that any investigation of the status of diagrams has to go beyond attempts at classifying them as sign-types and confront the question of their *necessity*. Why does one need them? What essential function — if any — do they serve? Could one do without them? Any answer depends, it seems, on who "one" is: mathematicians and scientists use them as abundantly and with as much abandon as those in the humanities avoid them. In fact, diagrams of any kind are so rare in the texts produced by historians, philosophers, and literary theorists, among others, that any instance sticks out like a sore thumb. An immediate response is to find this avoidance of visual devices totally unsurprising. Would not their embrace be stigmatized as scientism? Indeed, isn't the refusal to use figures, arrows, vectors, and so forth, as modes of explication part of the very basis on which the humanities define themselves as different from the technosciences? Why should texts committed — on whatever grounds — to communicating through words and not primarily interested in the sort of subject matter that lends itself to schematic visual representation make use of it? But this only pushes the question a little further out: What allows this prior commitment to words to be so self-sufficient, and what determines that certain topics or subjects, but not others, should lend themselves so readily to diagrammatic commentary and exegesis? Moreover, this separation by content isn't very convincing: philosophers are no less interested in space, time, and physical process than scientists; literary theorists occupy themselves quite as intensely as mathematicians with questions of pattern, analogy, opposition, and structure. Furthermore, whatever its value, such a response gives us no handle on the exceptions, the rare recourse to diagrams, that do occur in humanities texts.

To take a single example, how should we respond to the fact that in Husserl's entire oeuvre there is but one diagram (in his exegesis of temporality), a diagram that, interestingly enough, few commentators seem to make any satisfactory sense of? Are we to think that Husserl, trained as a mathematician, nodded — momentarily slipping from the philosophical into a more mathematical idiom? Or, unable to convey what he meant through words alone, did he resort, reluctantly perhaps but inescapably, to a picture? If the latter, then this fact — the possible inexplicability in words of his account of time — would surely be of interest in any overall analysis of Husserl's philosophical ideas. It would, after all, be an admission — highly significant in the present context — that the humanities' restriction to pictureless texts may be a warding off of uncongenial means of expression rather than any natural or intrinsic self-sufficiency in the face of its subject matter. But then would not this denial, or at any rate avoidance, of diagrams result in texts that were never free (at least never demonstrably so) of a willful inadequacy to their chosen exegetical and interpretive tasks; texts whose wordy opacity, hyper-elaboration, and frequent straining of written expression to the edge of sense were the reciprocal cost of this very avoidance?

And what goes for diagrams goes (with one exception) for ideograms: their absence is as graphically obvious as that of diagrams; the same texts in the humanities that avoid one avoid the other. And the result is an adherence to texts written wholly within the typographical medium of the alphabet. The exception is, of course, the writing of numbers: nobody, it seems, is prepared to dump the system that writes 7,654,321 in place of seven million, six hundred fifty-four thousand, three hundred twenty-one; the unwieldy prolixity here is too obvious to ignore. But why stop at numbers? Mathematics has many other ideograms and systems of writing — some of extraordinary richness and subtlety — besides the number notation based on o. What holds philosophers and textual theorists back? Although it doesn't answer this question, we can observe that the place-notation writing of numbers is in a sense a minimal departure from alphabetic typography: an ideogram like 7,654,321 being akin to a word spelled from the 'alphabet' 0,1,2,3,4,5,6,7,8,9 of 'letters', where to secure the analogy one would have to map the mathematical letter o onto something like a hyphen denoting the principled absence of any of the other given letters.

I alluded earlier to Weber's characterization of Peirce's semiotics as founded on a fear of ambiguity and the like. It's hard to resist seeing a reverse phobia in operation here: a recoil from ideograms (and, of course, diagrams) in the face of their potential to disrupt the familiar authority of the alphabetic text, an authority not captured but certainly anchored in writing's interpretability as the inscription of real or realizable speech. The apprehension and anxiety in the face of mathematical grams, which appear here in the form of writing as such — not as a recording of something prior to itself — are that they will always lead outside the arena of the speakable; one cannot, after all, *say* a triangle. If this is so, then the issue becomes the general relation among the thinkable, the writable, and the sayable, that is, what and how we imagine through different kinds of sign manipulations, and the question of their mutual translatability. In the case of mathematics, writing and thinking are cocreative and, outside the purposes of analysis and the like, impossible to separate.

Transferring the import of this from mathematics to spoken language allows one to see that speech, no less than mathematics, misunderstands its relation to the thinkable if it attempts a separation between the two into prior substance and posterior re-presentation; if, in other words, the form of an always re-presentational alphabetic writing is the medium through which speech articulates how and what it is. By withdrawing from the gram in this way, alphabetic writing achieves the closure of a false completeness, a self-sufficiency in which the fear of mathematical signs that motivates it is rendered as invisibly as the grams themselves. The idea of invisibility here, however, needs qualifying. Derrida's texts, for example, though written within and confined to a pure, diagramless and ideogramless format, nevertheless subvert the resulting alphabetic format and its automatic interpretation in terms of a vocalizable text through the use of various devices: thus a double text such as "Glas," which cannot be the inscription of any single or indeed dialogized speech, and his use of a neologism such as *différance*, which depends on and performs its meaning by being written and not said. But, all this notwithstanding, any attempt to pursue Harris's notion of a speechless link between the origin and the future of writing could hardly avoid facing the question of the meaning and use of diagrams. Certainly, Harris himself is alive to the importance and dangers of diagrams, as is evident from his witty taking apart of the particular diagram — a circuit of two heads speaking and hearing each other's thoughts — used by Saussure to illustrate his model of speech.[9]

But perhaps such a formulation, though it points in the right direction, is already — in light of contemporary developments — becoming inadequate. Might not the very seeing of mathematics in terms of a writing/thinking couple have become possible because writing is now — post-microchip — no longer what it was? I suggested above that the reflexivity of the relation between computing and mathematics — whereby the computer, having issued from mathematics, impinges on and ultimately transforms its originating matrix — might be the crux of the explanation for our late recognition of mathematics' status as writing.

To open up the point, I turn to a phenomenon within the ongoing microchip revolution, namely, the creation and implementation of what has come to be known as virtual reality. Although this might seem remote from the nature and practice of mathematics and from the issues that have so far concerned us, it is not, I hasten to add, *that* remote. In addition to many implicit connections to mathematical ideas and mathematically inspired syntax via computer programming, there are explicit links: thus, for example, Michael Benedikt, in his introductory survey of the historical and conceptual context of virtual reality, includes mathematics and its notations as an important thread running through the concept.[10]

An extrapolation of current practices more heralded, projected, and promised than as yet effectively realized, virtual reality comprises a range of effects and projects in which certain themes and practices recur. Thus, one always starts from the given world — the shared, intersubjective, everyday reality each of us inhabits. Within this reality is constructed a subworld, a space of virtual reality that we — or rather certain cyberneticized versions of ourselves — can, in some sense, enter and interact with. The construction of this virtuality, how it is realized — its parameters, horizons, possibilities, and manifestations — varies greatly from case to case. Likewise, what is entailed by a "version" of ourselves, and hence the sense in which "we" can be said to be "in" such virtual arenas, varies, since it will depend on what counts, for the purposes at hand, as physical immersion and interaction, and on how these are connected and eventually implemented. In all cases, however, virtual arenas are brought into existence inside computers and are entered and interacted with through appropriate interface devices and prosthetic extensions, such as specially adapted pointers, goggles, gloves, helmets, body sensors, and the like. Perhaps the most familiar example is dipping a single finger into a computer environment via the point-and-click operation of a computer mouse. But a mouse is a very rudimentary interface device, one which gives rise to a minimal interpretation, both in what the internalized finger can achieve as a finger and because a finger is, after all, only a metonym of a body: all current proposals call for more comprehensive prosthetics and richer, more fully integrated modes of interaction with/within these realities once they are entered.

Let's call the self in the world the default or *real-I*; the cyberneticized self we propel around a virtual world, the surrogate or *virtual-I*; and the self mediating between these, as the enabling site and means of their difference, the jacked-in or *goggled-I*. Operating a virtually real environment involves an interplay or circulation among these three agencies which ultimately changes the nature of the original, default reality, that is, of what it means to be a real-I inhabiting a/the given world. This circulation and especially its effect on, ultimately its transformation of, the given world motivate a great deal of virtual-reality thinking. To fix the point I'll mention two recent, differently conceived proposals, a social-engineering project and a fantasy extrapolation, which explore the possibilities offered by a virtualization of reality. The first, *Mirror Worlds*, is part propaganda, part blueprint for a vast series of public software projects by computer scientist David Gelernter, and the second, *Snow Crash*, is a science fiction epic of the near — cybered — future by novelist Neal Stephenson.[11]

Mirror Worlds sets itself the task of mapping out, more or less in terms of existing software technology, a way of virtualizing a public entity, such as a hospital or university or city (more ambitiously, an entire country, ultimately the world). Its aim is to create a virtual space, a computer simulation of, say, the city — what Gelernter calls the "agent space" — which each citizen could enter through various interface tools and engage in activities (education, shopping, information gathering, witnessing public events, monitoring and participating in cultural and political activity, meeting other citizens, and so on) in virtual form. The idea is that the results of such virtual-I activities would reflect back on society and effect changes in what it means to be a citizen within a community — to be a real-I — changes in previously unattainable and, given Gelernter's downbeat take on contemporary fragmentation and anomie, sorely needed ways. In *Snow Crash* Stephenson posits an America whose more computer-savvy denizens can move between a dystopian reality (panoptic surveillance and Mafia-franchised suburban enclaves) and a freely created, utopian computer space, the "metaverse," where their virtual-I's, or "avatars," can access the information net and converse and interact with each other in various virtual ways. Crucial to the plotting and thematics of Stephenson's narrative is the interplay between the inside of the metaverse and the all-too-real outside; the circulation, in other words, of affect and effects between virtual-I's and real-I's as the characters put on and take off their goggles. Although they move in opposite directions — in Stephenson's fantasy the virtual world in the end reflects the intrigue and violence of the real, while for Gelernter the virtual world is precisely the means of eliminating the anomic violence of the contemporary world — they share the idea of opposed worlds separated, joined, and mutually transformed by an interface.

A certain homology between virtual reality and mathematical thought, each organized around an analogous triad of agencies, should by now be evident. The virtual-I maps onto the mathematical Agent, the real-I onto the Person, and the goggled-I onto the Subject. In accordance with this mapping, both virtual reality and mathematics involve phenomenologically meaningful narratives of propelling a puppet — agent, simulacrum, surrogate, avatar, doppelganger, proxy (Peirce's "skeleton self") — of oneself around a virtual space. Both require a technology which gives real-I's access to this space and which controls the capabilities and characteristics of the skeleton-self agent. In both, this technology is structured and defined in terms of an operator, a figure with very particular and necessary features of its own, distinct from the puppet it controls and from the figure — the Person or real-I — occupying the default reality, able to put on goggles and operate in this way. And both are interactive in a material, embodied sense. In this they differ from the practices made possible by literature, which (like mathematics) conjures invisible proxies and identificatory surrogates of ourselves out of writing, and they differ from the media of theater, film, and TV, where (like virtual reality) proxies are not purely imagined, but have a visual presence. The difference arises from the fact that although these media allow, and indeed require of, their recipients/participants an active *interpretive* role, this doesn't and cannot extend to any real — materially effective — participation: mathematicians *manipulate* signs, and virtual realists *act out* journeys.

Mathematics, then, appears to be not only an enabling technology, but a template and precursor, perhaps the oldest one there is, of the current scenarios of virtual reality. But since something new is enabled here, what then (apart from obvious practical differences) distinguishes them? Surely, a principal and, in the present context, quite crucial difference lies in the instrumental means available to the operator-participants: the mathematical Subject's reliance on the writing technologies of ink and chalk inscriptions versus the prosthetic extensions available to the virtual reality operator. Therefore, what separates them is the degree of palpability they facilitate: the gap between the virtuality of a proxy whose repertoire (in the more ambitious projections) spans the entire sensorimotor range of modalities — ambulatory, auditory, proprioceptive, tactile, kinetic — and the invisible, disembodied Agent of mathematics. The virtual space entered by mathematicians' proxies is, in other words, entirely imagined, and the objects, points, functions, numbers, and so on, in it are without sensible form; percepts in the mind's eye rather than in the real eye necessary for virtual participation. Of course, the journeys that mathematical Agents perform, the narratives that can be told about them, the objects with which they react, and the regularities they encounter are strictly controlled by mathematical signs. Connecting these orders of signification, recreating the writing/thinking nexus through the interactive manipulation of visible diagrams and ideograms and the imagined, invisible states of affairs they signify and answer to, determines what it means to *do* mathematics.

We are thus led to the question: What if writing is no longer confined to inscriptions on paper and chalkboards, but becomes instead the creation of pixel arrangements on a computer screen? Wouldn't such a mutation in the material medium of mathematical writing effect a fundamental shift in what it means to think, and do, mathematics? One has only to bear in mind the changes in consciousness brought into play by the introduction of printing — surely a less radical conceptual and semiotic innovation than the shift from paper to screen — to think that indeed it would. The impact of screen-based visualization techniques on current scientific research and on the status of the theory/experiment opposition, as this has been traditionally formulated in the philosophy and history of science, already seems far-reaching. Thus, although primarily concerned with certain aspects of the recent computerization and mathematization of biology, the conclusion of Tim Lenoir and Christophe Lecuyer's investigation, namely, that "visualization *is* the theory," is suggestive far outside this domain.[12]

New types of mathematics — ways of thinking mathematically — have already come into existence precisely within the field of this mutation. Witness chaos theory and fractal geometry, with their essential reliance on computer-generated images (attractors in phase space, self-similar sets in the complex plane, and so on) which are nothing less than new, previously undrawable kinds of diagrams. And, somewhat differently, witness proofs (the four-color problem, classification of finite groups) that exist only as computer-generated entities. Moreover, there's no reason to suppose that this feedback from computer-created imagery and cognitive representations—in effect, a vector from an abstract, imagination-based technology to a concrete, image-based one—to the conceptual technology of mathematics will stop at the creation of new modes for drawing diagrams and notating arguments. But diagrams, because their meanings and possibilities stem from their genesis as physically drawn, bodily perceived objects, are already quasi-kinematic. In light of this, it's necessary to ask why such a process should be confined to the *visual* mode, to the creation of graphics and imagery, and not extended to the other sense modalities? What is to stop mathematics from appropriating the various computer-created ambulatory, kinesthetic, and tactile features made freely available within the currently proposed schemes for virtual reality? Is it unnatural or deviant to suggest that immersion in a virtually realized mathematical structure—walking around it, listening to it, moving and rearranging its parts, altering its shape, dismantling it, feeling it, and even smelling it, perhaps—be the basis for mathematical proofs? Would not such proofs, by using virtual experience as the basis for persuasion, add to, but go far beyond, the presently accepted practice of manipulating ideograms and diagrams in relation to an always invisible and impalpable structure? The understanding of writing appropriate to this conception of doing mathematics, what we might call *virtual writing*, would thus go beyond the "archewriting" set out by Derrida, since it could no longer be conceived in terms of the "gram" without wrenching that term out of all continuity with itself.

On at least one understanding of the genesis of thinking, nothing could be more natural and less deviant than using structures outside ourselves in order to think mathematics. Thus, according to Merlin Donald's recent account, the principal vector underlying the evolution of cognition and, ultimately, consciousness is the development and utilization of external forms of memory: our neuronal connections and hence our cognitive and imaginative capacities resulting, on this view, from forms of storage and organization outside our heads rather than the reverse.[13]

Evidently, natural or unnatural, such a transformation of mathematical practice would have a revolutionary impact on how we conceptualize mathematics, on what we imagine a mathematical object to be, on what we consider ourselves to be doing when we carry out mathematical investigations and persuade ourselves that certain assertions, certain properties and features of mathematical objects, are to be accepted as 'true'. Indeed, the very rules and protocols that control what is and isn't mathematically meaningful, what constitutes a 'theorem', for example, would undergo a sea change. An assertion would no longer have to be something capturable in a sentencelike piece of — presently conceived — writing, but could be a configuration that is meaningful only within a specifically presented virtual reality. Correspondingly, a proof would no longer have to be an argument organized around a written — as presently conceived — sequence of logically connected symbols, but could take on the character of an external, empirical verification. Mathematics would thus become what it has long denied being: an experimental subject; one which, though quite different from biology or physics in ways yet to be formulated, would be organized nonetheless around an independently existing, computer-created and -reproduced empirical reality.

This union, or rather this mutually reactive merging, of mathematics and virtual reality — a coming together of a rudimentary and yet-to-be consummated technology of manufactured presence with its ancient, highly developed precursor — would take the form of a double-sided process. As we've seen, from outside and independent of any mathematical desiderata, the goal of this technology is to achieve nothing less than the *virtualization of the real,* a process that will engender irreversible changes in what for us constitutes the given world, the default domain of the real-I. From the other side, in relation to a mathematics whose objects and structures have a wholly virtual, nonmaterial existence already, the process appears as the reverse, as effectively a *realization of the virtual,* whereby mathematical objects, by being constructed inside a computer, reveal themselves as materially presented and embodied, a process that will likewise cause irreversible and unexpected changes in the meaning we can attach to the idealized real. One could give a more specific content to this by looking at the most extravagantly virtual concept of contemporary mathematics, that of infinity; a concept so inherently metaphysical and spectral as to be unrealizable — actually or even in principle — within the universe we inhabit. Or so I have argued at length elsewhere.[14]

Of course, these remarks on the joint future of mathematics and virtual reality, although they could be supported by looking at computer-inflected practices within the current mathematical scene, are little more than highly speculative extrapolation here. I have included them in order to get a fresh purchase on the notion of a diagram: by overtly generalizing the notion into a virtually realized (i.e., physically presented) mathematical structure, one can see how the question of diagrams is really a question of the body. To exclude diagrams — either deliberately as part of the imposition of rigor in mathematics or less explicitly as an element in a general and unexamined refusal to move beyond alphabetic texts and the linear strings of ideograms that mimic them within mathematics — is to occlude materiality, embodiment, and corporeality, and hence the immersion in history and the social that is both the condition for the possibility of signifying and its (moving) horizon.

NOTES

This article was written during the period when I was supported by a fellowship from the National Endowment for the Humanities.

1   Roy Harris, *The Origin of Writing* (London, 1986), 158.

2   Edmund Husserl, "The Origin of Geometry," trans. David Carr, in *Husserl: Shorter Works*, ed. P. McCormick and F. Ellison (Notre Dame, 1981), 257.

3   Brian Rotman, "Toward a Semiotics of Mathematics," *Semiotica* 72 (1988): 1–35; and *Ad Infinitum . . . The Ghost in Turing's Machine: Taking God Out of Mathematics and Putting the Body Back In* (Stanford, 1993), 63–113.

4   Charles Sanders Peirce, *Collected Writings*, ed. Philip Weiner (New York, 1958), 5: 332.

5   Samuel Weber, *Institution and Interpretation* (Minneapolis, 1987), 30.

6   Maurice Merleau-Ponty, *The Phenomenology of Perception*, trans. Colin Smith (London, 1962), 388.

7   Philip Davis, "Visual Theorems," *Educational Studies in Mathematics* 24 (1993): 333–44.

8   See Brian Rotman, *Signifying Nothing: The Semiotics of Zero* (Stanford, 1993).

9   Roy Harris, *The Language Machine* (Ithaca, 1987), 149–52.

10 Michael Benedikt, *Cyberspace: First Steps* (Cambridge, 1992), 18–22.

11 David Gelernter, *Mirror Worlds* (New York, 1992); Neal Stephenson, *Snow Crash* (New York, 1992).

12 Timothy Lenoir and Christophe Lecuyer, "Visions of Theory" (to appear).

13 Merlin Donald, *Origins of the Modern Mind* (Cambridge, MA, 1991).

14 Rotman, *Ad Infinitum*.

Concepts and the Mangle of Practice:

Constructing Quaternions

Andrew Pickering

Similarly, by surrounding $\sqrt{-1}$ by talk about vectors, it sounds quite natural to talk of a thing whose square is -1. That which at first seemed out of the question, if you surround it by the right kind of intermediate cases, becomes the most natural thing possible. — Ludwig Wittgenstein, *Lectures on the Foundations of Mathematics*

How can the workings of the mind lead the mind itself into problems? . . . How can the mind, by methodical research, furnish itself with difficult problems to solve? . . . This happens whenever a definite method meets its own limit (and this happens, of course, to a certain extent, by chance). — Simone Weil, *Lectures on Philosophy*

AN asymmetry exists in our accounts of scientific practice: machines are located in a field of agency, but concepts are not.[1] Thus while it is easy to appreciate that dialectics of resistance and accommodation can arise in our dealings with machines (I have argued elsewhere that the contours of material agency emerge only in practice[2]), it is hard to see how the same can be said of our dealings with concepts. And, this being the case, the question arises of why concepts are not mere putty in our hands. Why is conceptual practice difficult? "How can the workings of the mind lead the mind itself into problems?" It seems to me that one cannot claim to have a full analysis of scientific practice until one can suggest answers to questions like these, and in this paper I argue, first in the abstract then via an example, that a symmetrizing move is needed. We should think of conceptual structures as themselves located in fields of agency, and of the transformation and extension of such structures as emerging in dialectics of resistance and accommodation within those fields — dialectics which, for short, I call *the mangle of practice*.

When we think, we are conscious that a connection between feelings is determined by a general rule, we are aware of being governed by a habit. Intellectual power is nothing but facility in taking habits and in following them in cases essentially analogous to, but in non-essentials widely remote from, the normal cases of connections of feelings under which those habits were formed. — Charles Sanders Peirce, *Chance, Love and Logic*

The student of mathematics often finds it hard to throw off the uncomfortable feeling that his science, in the person of his pencil, surpasses him in intelligence. — Ernst Mach, quoted by Ernest Nagel, *Teleology Revisited*

My analysis of conceptual practice depends upon and elaborates three central ideas: first, that cultural practices (in the plural) are disciplined and machinelike; second, that practice, as cultural extension, is centrally a process of open-ended modeling; and third, that modeling takes place in a field of cultural multiplicity and is oriented to the production of associations between diverse cultural elements. I can take these ideas in turn.

Think of an established conceptual practice — elementary algebra, say. To know algebra is to recognize a set of characteristic symbols *and how to use them*. As Wittgenstein put it: "Every sign *by itself* seems dead. *What* gives it life? — In use it is *alive*."[3] And such uses are disciplined; they are machinelike actions, in Harry Collins's terminology.[4] Just as in arithmetic one completes "3 + 4 =" by writing "7" without hesitation, so in algebra one automatically multiplies out "$a(b + c)$" as "$ab + ac$." Conceptual systems, then, hang together with specific, disciplined patterns of human agency, particular routinized ways of connecting marks and symbols with one another. Such disciplines — acquired in training and refined in use — carry human conceptual practices along, as it were, independently of individual wishes and intents. The scientist is, in this sense, passive in disciplined conceptual practice. This is a key point in what follows, and, in order to mark it and to symmetrize the formulation, I want to redescribe such human passivity in terms of a notion of *disciplinary agency*. It is, I shall say, the agency of a discipline (elementary algebra, for example) that leads disciplined practitioners through a series of manipulations within an established conceptual system.[5]

I will return to disciplinary agency in a moment, but now we can turn from disciplined practices to the practice of cultural extension. A point that I take to be established about conceptual practice is that it proceeds through a process of modeling. Just as new machines are modeled on old ones, so are new conceptual structures modeled upon their forebears.[6] And much of what follows takes the form of a decomposition of the notion of modeling into more primitive elements. As it appears in my example, at least, it is useful to distinguish three stages within any given modeling sequence. Briefly, modeling has to be understood, I think, as an open-ended process having no predetermined destination, and this is certainly true of conceptual practice. Part of modeling is thus what I call *bridging*, or the construction of a *bridgehead* that tentatively fixes a vector of cultural extension to be explored. Bridging, however, is not sufficient to efface the openness of modeling; it is not enough in itself to define a new conceptual system on the basis of an old one. Instead, it marks out a space for *transcription* — the copying of established moves from the old system into the new space fixed by the bridgehead (hence my use of the word "bridgehead"). And, if my example is a reliable guide, even transcription can be insufficient to complete the modeling process. What remains is *filling*, completing the new system in the absence of any clear guidance from the base model.

Now, this decomposition of modeling into bridging, transcription, and filling is at the heart of my analysis of conceptual practice, and I will clarify what these terms mean when we come to the example. For the moment, though, I want to make a general remark about how they connect to issues of agency. As I conceive them, bridging and filling are activities in which scientists display choice and discretion, the classic attributes of human agency. Scientists are active in these phases of the modeling process, in Fleck's sense.[7] Bridging and filling are *free moves*, as I shall say. In contrast, transcription is where discipline asserts itself, where disciplinary agency carries scientists along, where scientists become passive in the face of their training and established procedures. Transcriptions, in this sense, are disciplined *forced moves*. Conceptual practice therefore has, in fact, the form of a *dance of agency* in which the partners are alternately the classic human agent and disciplinary agency. And two points are worth emphasizing here. First, this dance of agency, which manifests itself at the human end in the intertwining of free and forced moves in practice, is not optional. Practice has to take this form. The *point* of bridging as a free move is to invoke the forced moves that follow from it. Without such invocation, conceptual practice would be empty. Second, the intertwining of free and forced moves implies what Yves Gingras and S. S. Schweber refer to (rather misleadingly) as a certain "rigidity" of conceptual "networks."[8] I take this reference as a gesture toward the fact that scientists are not fully in control of where passages of conceptual practice will lead. Conceptual structures, one can say, relate to disciplinary agency much as machines do to material agency. Once one begins to tinker with the former, just as with the latter, one has to find out in practice how the resulting conceptual machinery will perform. It is precisely in this respect that dialectics of resistance and accommodation can arise in conceptual practice. To see how, though, requires some further discussion.

The constitutive role of disciplinary agency in conceptual practice is enough to guarantee that its endpoints are temporally emergent. One simply has to play through the moves that follow from the construction of specific bridgeheads and see where they lead. But this is not enough to explain the emergence of resistance, to get at how "the workings of the mind lead the mind itself into problems." To get at this, one needs to understand what conceptual practice is for. I do not suppose that any brief, general answer to this question exists, but all of the examples that I can think of lead to themes of cultural multiplicity and the making and breaking of associations between diverse cultural elements. Some general remarks on empirical science and mathematics can illustrate what I have in mind.

In science, one prominent object of conceptual practice is bringing theoretical ideas to bear upon empirical data to understand or explain the latter, to extract supposedly more fundamental information from them, or whatever. In *Constructing Quarks,* I argued that this process was indeed one of modeling, and now I would add four remarks. First, this process points to the multiplicity (and heterogeneity) of scientific culture. Data and theory have no necessary connection to one another; such connections as exist between them have to be made. Hence my second point: conceptual practice aims at making associations (trans- lations, alignments) between such diverse elements—here, data and theory. Third, just because of the presence of the disciplinary partner in the dance of agency in conceptual practice, resistances can arise in the making of such associations. Because the endpoints of conceptual practice cannot be known in advance, the pieces do not necessarily fit together as intended. And fourth, these resistances precipitate dialec- tics of resistance and accommodation, tentative revisions of modeling vectors, manglings that can bear upon conceptual structures as well as the form and performance of material apparatus.[9]

To exemplify these ideas, an obvious strategy would be to document how disciplines structure practice in theoretical science, but I will not take that route here because the disciplines and conceptual structures at stake in all of the interesting cases that I know about—largely in recent theoretical elementary-particle physics—are sufficiently esoteric to make analysis and exposition quite daunting. I propose instead to concentrate on mathematics, and in particular on an example from the history of mathematics which is at once intellectually and historically interesting and simple, drawing only upon relatively low-level and al- ready familiar disciplines and structures in basic algebra and geometry. I hope thus to find an example of the mangle in action in conceptual practice that is accessible while being rich enough to point to exten-sions of the analysis, in science proper as well as in mathematics. I will come to the example in a moment, but first some remarks are needed on mathematics in general.

Physics might be said to seek, among other things, somehow to describe the world, but what is mathematics for? Once again, I sup-pose that there is no general answer to this question, but I think Latour makes some important and insightful moves. In his discussion of mathematical formalisms, Latour continually invokes metaphors of joining, linking, association, and alignment, comparing mathematical structures to railway turntables, to crossroads and cloverleaf junctions, and to telephone exchanges.[10] His idea, then, is that such structures themselves serve as multipurpose translation devices, making connec-tions among diverse cultural elements. And, as we shall see, this turns out to be the case in our example. The details follow, but the general point can be made in advance. If cultural extension in conceptual prac-tice is not fully under the control of active human agents due to the constitutive role of disciplinary agency, then the making of new as-sociations — the construction of new telephone exchanges linking new kinds of subscribers — is nontrivial. Novel conceptual structures need to be tuned if they are to stand a chance of performing cooperatively in fields of disciplinary agency; one has to expect that resistances will arise in the construction of new conceptual associations, precipitating continuing dialectics of resistance and accommodation, manglings of modeling vectors — of bridgeheads and fillings, and even of disciplines themselves.[11]

This is the process that we can now explore in an example taken from the history of mathematics. In the next two sections we will be concerned with the work of the great Irish mathematician Sir William Rowan Hamilton, particularly with a brief passage of his mathemati-cal practice that culminated on 16 October 1843 in the construction of his new mathematical system of quaternions. First, however, I want to remark on the selection of this example, which recommends itself on several counts. The disciplinary agency manifested in Hamilton's work has a simple and familiar structure, thus making his work much easier to follow than that of present-day mathematicians or scientists. At the same time, Hamilton's achievement in constructing quaternions is of considerable historical interest. It marked a turning point in the development of mathematics, involving as it did the introduction of noncommuting quantities into the subject matter of the field as well as an exemplary set of new entities and operations, the quaternion system, that mutated over time into the vector analysis central to modern physics. And further, detailed documentation of Hamilton's practice is available.[12] Hamilton himself left several accounts of the passage of practice that led him to quaternions, especially a notebook entry written on the day of the discovery and a letter to John T. Graves dated the following day.[13] As Hamilton's biographer put it, "These documents make the moment of truth on Dublin Bridge" (where Hamilton first conceived of the quaternion system) "one of the best-documented discoveries in the history of mathematics."[14]

Hamilton's discovery of quaternions is not just well-documented, it is also much written about. Most accounts of Hamilton's algebraic researches contain some treatment of quaternions, and at least five accounts in the secondary literature rehearse to various ends Hamilton's own accounts, more or less in their entirety.[15] What differentiates my account from others is my desire to show that Hamilton's work can be grasped within the more general understanding of agency and practice that I call the mangle. Together with the notions of free and forced moves and disciplinary agency, that of the open-endedness of modeling is especially important here, and in what follows I seek to locate the free moves in Hamilton's eventual route to quaternions by setting that trajectory in relation to his earlier attempts to construct systems of "triplets."

The early nineteenth century was a time of crisis in the foundation of algebra, centering on the question of how the "absurd" quantities — negative numbers and their square roots — should be understood.[16] Various moves were made in the debate over the absurd quantities, only one of which bears upon our story and which serves to introduce the themes of cultural multiplicity and association as they will figure here. This was the move to construct an association between algebra and an otherwise disparate branch of mathematics — geometry — an association that consisted in establishing a *one-to-one correspondence* between the elements and operations of complex algebra and a particular geometrical system.[17] I need to go into some detail about the substance of this association, since it figured importantly in Hamilton's construction of quaternions.[18]

The standard algebraic notation for a complex number is $x + iy$, where x and y are real numbers and $i^2 = -1$. Positive real numbers can be thought of as representing measurable quantities or magnitudes — a number of apples, the length of a rod — and the foundational problem in algebra was to think what -1 and i (and multiples thereof) might

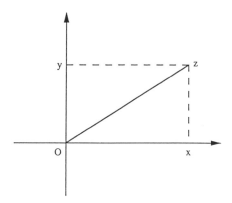

Figure 1a. Geometrical representation of the complex number $z = x + iy$ in the complex plane. The projections onto the x- and y-axes of the endpoint of the line Oz measure the real and imaginary parts of z, respectively.

stand for. What sense can one make of $\sqrt{-1}$ apples? How many apples is that? The geometrical response to such questions was to think of x and y not as quantities or magnitudes, but as coordinates of the endpoint of a line-segment terminating at the origin in some "complex" two-dimensional plane. Thus the x-axis of the plane measured the real component of a given complex number represented by such a line-segment, and the y-axis the imaginary part, the part multiplied by i in the algebraic expression (Figure 1a). In this way the entities of complex algebra were set in a one-to-one correspondence with geometrical line-segments. Further, it was possible to put the operations of complex algebra in a similar relation with suitably defined operations upon line-segments. Addition of line-segments was readily defined on this criterion. In algebraic notation, addition of two complex numbers was defined as

$$(a + ib) + (c + id) = (a + c) + i(b + d),$$

and the corresponding rule for line-segments was that the x-coordinate of the sum should be the sum of the x-coordinates of the segments to be summed, and likewise for the y-coordinate (Figure 1b). The rule for subtraction could be obtained directly from the rule for addition — coordinates of line-segments were to be subtracted instead of summed.

The rules for multiplication and division in the geometrical representation were more complicated, but we need only discuss that for multiplication, since this was the operation that became central to Hamilton's development of quaternions. The rule for algebraic multiplication of two complex numbers,

$$(a + ib)(c + id) = (ac - bd) + i(ad + bc),$$

followed from the usual rules of algebra, coupled with the peculiar definition of $i^2 = -1$. The problem was then to think what the equivalent might be in the geometrical representation. It proved to be stateable as the conjunction of two rules: the product of two line-segments is another line-segment that (a) has a length given by the product of the lengths of the two segments to be multiplied and that (b) makes an angle with the x-axis equal to the sum of the angles made by the two segments (Figure 2). From this definition, it is easy to check that multiplication of line-segments in the geometrical representation gives a result equivalent to the multiplication of the corresponding complex numbers in the algebraic representation.[19] Coupled with a suitably contrived definition of division in the geometrical representation, then, an association of one-to-one correspondence was achieved between the entities and operations of complex algebra and their geometrical representation in terms of line-segments in the complex plane.

At least three important consequences for nineteenth-century mathe-

Figure 1b. Addition of complex numbers in the geometrical representation: $z_3 = z_1 + z_2$. By construction, the real part of z_3 is the sum of the real parts of z_1 and z_2 ($x_3 = x_1 + x_2$), and likewise the imaginary part ($y_3 = y_1 + y_2$).

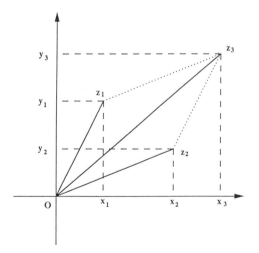

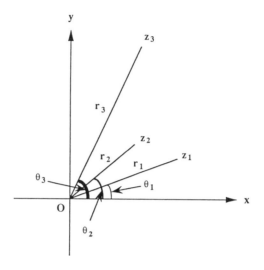

Figure 2. Multiplication of complex numbers in the geometrical representation: $z_3 = z_1 \times z_2$. Here the lengths of line-segments are multiplied ($r_3 = r_1 \times r_2$), while the angles subtended with the x-axis by line-segments are added ($\theta_3 = \theta_1 + \theta_2$).

matics flowed from this association. First, it could be said (though it could also be disputed) that the association solved the foundational problems centered on the absurd numbers. Instead of trying to understand negative and imaginary numbers as somehow measures of quantities or magnitudes of real objects, one should think of them geometrically, in terms of the orientation of line-segments. A negative number, for example, should be understood as referring to a line-segment lying along the negative (rather than the positive) x-axis (Figure 3a), a pure imaginary number as lying along the y-axis (Figure 3b), and so on. Thus for an understanding of the absurd numbers one could appeal to an intuition of the possible differences in length and orientation of rigid bodies — sticks, say — in any given plane, and hence the foundational problem could be shown to be imaginary rather than real (so to speak).

Second, more practically, the geometrical representation of complex algebra functioned as a switchyard. Algebraic problems could be reformulated as geometrical ones, and thus perhaps solved using geometric techniques, and vice versa. The third consequence of this association of algebra with geometry was that the latter, more clearly than the former, invited extension. Complex algebra was a self-contained field of mathematical practice; geometry, in contrast, was by no means confined to the

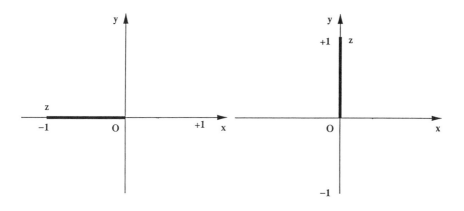

Figure 3a. Geometrical representations of absurd numbers: (a) $z = -1$.

plane. The invitation, then, was to extend the geometrical representation of complex-number theory from a two- to a three-dimensional space and to somehow carry along with it a three-place algebraic equivalent, maintaining the association already constructed in two dimensions. On the one hand, this extension could be attempted in a spirit of play, just to see what could be achieved; on the other, there was a promise of utility. The hope was to construct an algebraic replica of transformations of line-segments in three-dimensional space, and thus to develop a new and possibly useful algebraic system appropriate to calculations in three-dimensional geometry, "to connect, in some new and useful (or at least interesting) way, *calculation* with *geometry*, through some *extension* [of the association achieved in two dimensions], to *space of three dimensions*," as Hamilton put it.[20]

Hamilton was involved in the development of complex algebra from the late 1820s on. He worked both on the foundational problems just discussed (developing his own approach to them via his "Science of Pure Time" and a system of "couples" rather than through geometry; I will return to this topic later) and on the extension of complex numbers from two- to three-place systems, or "triplets," as he called them. His attempts to construct triplet systems in the 1830s were many and various, but Hamilton regarded them all as failures.[21] Then, in 1843, after a period of work on other topics, he returned to the challenge once more. Again, he failed to achieve his goal, but this time he did not come away empty-handed. Instead of constructing a three-place or three-dimensional system, he quickly arrived at the four-place quater-

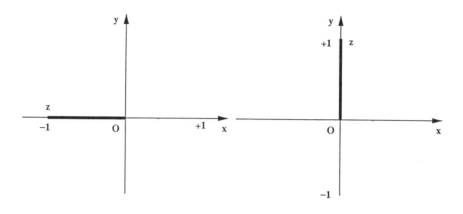

Figure 3b. $z = i = \sqrt{-1}$.

nion system that he regarded as his greatest mathematical achievement
and to which he devoted the remainder of his life's work.

On 16 October 1843, Hamilton set down in a notebook his recollection
of his path to quaternions. The entry begins:

> I, this morning, was led to what seems to me a theory of *qua-*
> *ternions,* which may have interesting developments. *Couples* being
> supposed known, and known to be representable by points in a
> plane, so that $\sqrt{-1}$ is perpendicular to 1, it is natural to conceive
> that there may be another sort of $\sqrt{-1}$, perpendicular to the plane
> itself. Let this new imaginary be *j;* so that $j^2 = -1$, as well as $i^2 = -1$.
> A point *x, y, z* in space may suggest the triplet $x + iy + jz$.[22]

I can begin my commentary on this passage by noting that a process
of modeling was constitutive of Hamilton's practice. As is evident from
these opening sentences, he did not attempt to construct a three-place
mathematical system out of nothing. Instead, he sought to move into
the unknown from the known, to find a creative extension of the two-
place systems already in existence. Further, as will become evident as
we go along, the process of cultural extension through modeling was,
in this instance as in general, an open-ended one: in his work on triplet
systems that culminated in the construction of quaternions, Hamilton
tried out a large number of different extensions of complex algebra and
geometry. Now I need to talk about how Hamilton moved around in

this open-ended space, a discussion that will lead us into the crucial tripartite decomposition of modeling mentioned earlier.

In his reference to "points in a plane," Hamilton first invoked the geometrical representation of complex algebra, and the extension that he considered was to move from thinking about line-segments in a plane to thinking about line-segments in a three-dimensional space. In so doing, he established what I call a *bridgehead* to a possible three-dimensional extension of complex algebra. The significance of such a bridging operation is that it marks a particular destination for modeling, and here I want to emphasize two points that I suspect are general about bridging. First, however natural Hamilton's specific move from the plane to three-dimensional space may seem, it is important to recognize that it was by no means forced upon him. In fact, in his earlier attempts at triplet systems, he had proceeded differently, often working first in terms of an algebraic model and only toward the end of his calculations seeking geometrical representations of his findings, representations which were quite dissimilar from that with which he began here.[23] In this sense, the act of fixing a bridgehead is an active or free move that serves to cut down the indefinite openness of modeling. My second point follows from this: such free moves need to be seen as tentative and revisable trials that carry no guarantee of success. Just as Hamilton's earlier choices of bridgeheads had, in his own estimation, led to failure, so might this one. His only way of assessing this particular choice was to work with it and on it — to see what he could make of it. Similar comments apply to the second model that structured Hamilton's practice — the standard algebraic formulation of complex numbers — which he extended (in the quotation above), to a three-place system by moving from the usual $x + iy$ notation to $x + iy + jz$. This seems like another natural move to have made. But again, when set against Hamilton's earlier work on triplets, it is better seen as the establishment of a bridgehead in a tentative free move.[24]

One more remark before returning to Hamilton's recollections: I noted above that complex algebra and its geometrical representation were associated with one another in a relation of one-to-one correspondence, and an intent to preserve that association characterized the passage of Hamilton's practice under discussion here. The excerpt quoted above shows that he set up a one-to-one correspondence between the *elements* defined in his two bridging moves — between the algebraic notation $x + iy + jz$ and suitably defined three-dimensional line-segments. The next passage shows that he considered the possibility of preserving the same association of mathematical *operations* in the two systems. This is where the analysis of modeling becomes interesting, where disciplinary agency comes into play and the possibility of resistance in conceptual practice thus becomes manifest. Hamilton's notebook entry continues:

> The square of this triplet $[x + iy + jz]$ is on the one hand $x^2 - y^2 - z^2 + 2ixy + 2jxz + 2ijyz$; such at least it seemed to me at first, because I assumed $ij = ji$. On the other hand, if this is to represent the third proportional to 1, 0, 0 and x, y, z, considered as *indicators of lines*, (namely the lines which end in the points having these coordinates, while they begin at the origin) and if this third proportional be supposed to have its length a third proportional to 1 and $\sqrt{(x^2 + y^2 + z^2)}$, and its distance twice as far angularly removed from 1, 0, 0 as x, y, z; then its real part ought to be $x^2 - y^2 - z^2$ and its two imaginary parts ought to have for coefficients $2xy$ and $2xz$; thus the term $2ijyz$ appeared de trop, and I was led to assume at first $ij = 0$. However I saw that this difficulty would be removed by supposing that $ji = -ij$.[25]

This passage requires some exegesis. Here, Hamilton began to think about mathematical operations on the three-place elements that his bridgeheads had defined, and in particular about the operation of multiplication, specialized initially to that of squaring an arbitrary triplet. He worked first in the purely algebraic representation; writing $t = x + iy + jz$, he found:

$$t^2 = x^2 - y^2 - z^2 + 2ixy + 2jxz + 2ijyz. \tag{1}$$

This equation follows automatically from the laws of standard algebra, coupled with the usual definition that $i^2 = -1$ and the new definition $j^2 = -1$ that was part of Hamilton's algebraic bridgehead. In this instance, then, we see that the primitive notion of modeling can be partly decomposed into two, more transparent operations, bridging and *transcription*, where the latter amounts to copying an operation defined in the base model—in this instance, the rules of algebraic multiplication—into the system set up by the bridgehead. And this, indeed, is why I use the word "bridgehead": it defines a point to which attributes of the base model can be transferred, a destination for modeling, as I put it earlier. We can note here that just as it is appropriate to think of fixing a bridgehead as an active, free move, it is likewise appropriate to think of transcription as a sequence of passive, forced moves, a sequence of moves— resulting here in equation (1)—that follows from what is already established concerning the base model. And we can note, further, that the surrender of agency on Hamilton's part is equivalent to the assumption of agency by discipline. While Hamilton was indeed the person who thought through and wrote out the multiplications in question, he was not free to choose how to perform them. Anyone already disciplined in algebraic practice, then or now, can check that Hamilton (and I) did the multiplication correctly. This, then, is our first example of the dance of agency in conceptual practice, in which disciplinary agency carried Hamilton (and carries us) beyond the fixing of a bridgehead.

The disciplined nature of transcription is what makes possible the emergence of resistance in conceptual practice, but before we come to that we should note that the decomposition of modeling into bridging and transcription is only partial. Equation (1) still contains an undefined quantity—the product ij—that appears in the last term of the right-hand side. This was determined neither in Hamilton's first free move nor in the forced moves that followed. The emergence of such "gaps" is, I believe, another general feature of the modeling process: disciplinary agency is insufficient to carry through the processes of cultural extension that begin with bridging. Gaps appear throughout Hamilton's work on triplets, for example, and one typical response of his was what I call *filling*, meaning the assignment of values to undefined terms in further free moves.[26] Retaking the lead in the dance of agency, Hamilton could have, say, simply assigned a value to the product ij and explored where that led him through further forced moves. In this instance, though, he proceeded differently.

The sentence that begins "On the other hand, if this is to represent the third proportional" refers to the operation of squaring a triplet in the geometrical rather than the algebraic representation. Considering a triplet as a line-segment in space, Hamilton was almost in a position to transcribe onto his new bridgehead the rules for complex multiplication summarized above, but, although not made explicit in the quoted passage, one problem remained. While the first rule, concerning the length of the product of lines, remained unambiguous in three-dimensional space, the second, concerning the orientation of the product line, did not. Taken literally, it implied that the angle made by the square of any triplet with the x-axis was twice the angle made by the triplet itself— "twice as far angularly removed from 1, 0, 0 as x, y, z"—but it in no way specified the orientation of the product line in space. Here, disciplinary agency again left Hamilton in the lurch. Another gap thus arose in moving from two to three dimensions, and, in this instance, Hamilton responded with a characteristic, if unacknowledged, filling move.

He further specified the rule for multiplication of line-segments in space by enforcing the new requirement that the square of a triplet remain in the plane defined by itself and the x-axis. (This is the only way in which one can obtain his stated result for the square of a triplet in the geometrical representation.) As usual, this move seems natural enough, but the sense of naturalness is readily disturbed when taken in the context of Hamilton's prior practice. One of Hamilton's earliest attempts at triplets, for example, represented them as lines in three-dimensional space, but multiplication was defined differently in that attempt.[27] Be that as it may, this particular filling move sufficed and was designed to make possible a series of forced transcriptions from the two- to the three-dimensional version of complex algebra that enabled Hamilton to compute the square of an arbitrary triplet. Surrendering once more to the flow of discipline, he found that the "real part" of the corresponding line-segment "ought to be $x^2 - y^2 - z^2$ and its two imaginary parts ought to have for coefficients $2xy$ and $2xz$." Or, returning this result to purely algebraic notation,[28]

$$t^2 = x^2 - y^2 - z^2 + 2ixy + 2jxz. \tag{2}$$

Now, there is a simple difference between equations (1) and (2), both of which represent the square of a triplet but calculated in different ways. The two equations are identical except that the problematic term $2ijyz$ of equation (1) is absent from equation (2). This, of course, is just the kind of thing that Hamilton was looking for to help him in defining the product ij, and we will examine the use he made of it in a moment. First, it is time to talk about *resistance*. The two base models that Hamilton took as his points of departure — the algebraic and geometrical representations of complex numbers — were associated in a one-to-one correspondence of elements and operations. Here, however, we see that, as so far extended by Hamilton, the three-place systems had lost this association. The definition of a square in the algebraic system (equation [1]) differed from that computed via the geometrical representation (equation [2]). The association of "calculation with geometry" that Hamilton wanted to preserve had been broken; a resistance to the achievement of Hamilton's goal had appeared. And, as I have already suggested, the precondition for the emergence of this resistance was the constitutive role of disciplinary agency in conceptual practice and the consequent intertwining of free and forced moves in the modeling process. Hamilton's free moves had determined the directions that his extensions of algebra and geometry would take in the indefinitely open space of modeling, but the forced moves intertwined with them had carried those extensions along to the point where they collided in equations (1) and (2). This, I think, is how "the workings of the mind lead the mind itself into problems." We can now move from resistance itself to a consideration of the dialectic of resistance and accommodation in conceptual practice, in other words to the mangle.

The resistance that Hamilton encountered in the disparity between equations (1) and (2) can be thought of as an instance of a generalized version of the Duhem problem.[29] Something had gone wrong somewhere in the process of cultural extension — the pieces did not fit together as desired — but Hamilton had no principled way of knowing where. What remained for him to do was to tinker with the various extensions in question — with the various free moves he had made, and thus with the sequences of forced moves that followed from them — in the hope of getting around the resistance that had arisen and achieving the desired association of algebra and geometry. He was left with seeking some *accommodation* to resistance. Two possible starts toward accommodation are indicated in the passage last quoted, both of which amounted to further fillings-in of Hamilton's extended algebraic system and both of which led directly to an equivalence between equations (1) and (2). The most straightforward accommodation was to set the product ij equal to zero.[30] An alternative, less restrictive but more dramatic and eventually more far-reaching move also struck Hamilton as possible, namely, to abandon the assumption of commutation between i and the new square root of -1, j.[31] In ordinary algebra, this assumption — which is to say that $ab = ba$ — was routine. Hamilton entertained instead the possibility that $ij = -ji$. This did not rule out the possibility that both ij and ji were zero; but even without this being the case, it did guarantee that the problematic term $2ijyz$ of equation (1) vanished, and it thus constituted a successful accommodation to the resistance that had emerged at this stage.[32]

Hamilton thus satisfied himself that he could maintain the association between his algebraic and geometrical three-place systems by the assumption that i and j did not commute, at least as far as the operation of squaring a triplet was concerned. His next move was to consider a less restrictive version of the general operation of multiplication, working through, as above, the operation of multiplying two coplanar but otherwise arbitrary triplets. Again, he found that the results of the calculation were the same in the algebraic and geometrical representations as long as he assumed either $ij = 0$ or $ij = -ji$.[33] Hamilton then moved on to consider the fully general instance of multiplication in the new formalism, the multiplication of two arbitrary triplets.[34] As before, he began in the algebraic representation. Continuing to assume $ij = -ji$, he wrote:

$$(a + ib + jc)(x + iy + jz) = ax - by - cz + i(ay + bx) \\ + j(az + cx) + ij(bz - cy). \qquad (3)$$

He then turned back to thinking about multiplication within the geometrical representation, where a further problem arose. Recall that in defining the operation of squaring a triplet Hamilton had found it necessary to make a filling free move, assuming that the square lay in the plane of the original triplet and the x-axis. This filling move was sufficient to lead him through a series of forced moves to the calculation of the product of two arbitrary but coplanar triplets. It was insufficient, however, to define the orientation in space of the product of two completely arbitrary triplets: in general, one could not pass a plane through any two triplets and the x-axis. Once more, Hamilton could have attempted a filling move here, concocting some rule for the orientation of the product line in space, say, and continuing to apply the sum rule for the angle made by the product with the x-axis. In this instance, however, he followed a different strategy.

Instead of attempting the transcription of the two rules that fully specified multiplication in the standard geometrical representation of complex algebra, he began to work only in terms of the first rule — that the length of the product line-segment should be the product of the lengths of the line-segments to be multiplied. Transcribing this rule to three dimensions, and working for convenience with squares of lengths, or "square moduli," rather than with lengths themselves, he could surrender his agency to Pythagoras's theorem and write the square modulus of the left-hand side of equation (3) as $(a^2 + b^2 + c^2)(x^2 + y^2 + z^2)$ — another forced move.[35] Now he had to compute the square of the length of the right-hand side. Here the obstacle to the application of Pythagoras's theorem was the quantity ij again appearing in the last term. If Hamilton assumed that $ij = 0$, the theorem could be straightforwardly applied, giving a value for the square modulus of $(ax - by - cz)^2 + (ay + bx)^2 + (az + cx)^2$. The question now was whether these two expressions for the lengths of the line-segments appearing on the two sides of equation (3) were equal. Hamilton multiplied them out and rearranged the expression for the square modulus of the left-hand side, finding that it in fact differed from that on the right-hand side by a factor of $(bz - cy)^2$. Once again, a resistance had arisen, here in thinking about the product of two arbitrary triplets in, alternatively, the algebraic and the geometrical representation. Once more, the two representations, extended from two- to three-place systems, led to different results. And once more, Hamilton looked for some accommodation to this resistance, for some way of making the two notions of multiplication equivalent, as they were in two dimensions.

The new resistance was conditional on the assumption that $ij = 0$. The question, then, was whether some other assignment of ij might succeed in balancing the moduli of the left- and right-hand sides of equation (3).[36] And here Hamilton made a key observation: the superfluous term in the square modulus of the left-hand side of equation (3), $(bz - cy)^2$, was the square of the coefficient of ij on the right-hand side. The two computations of the square modulus could thus be made to balance by assuming not that the product of i and j vanished, but that it was some third quantity k, a *"new imaginary,"* different again from i and j such that Pythagoras's theorem could be applied to it, too.[37]

The introduction of the new imaginary k, defined as the product of i and j, thus constituted a further accommodation by Hamilton to an emergent resistance in thinking about the product of two arbitrary triplets in terms of both algebraic and geometrical representations at once, and one aspect of this particular accommodation is worth emphasizing: It amounted to a drastic shift of bridgehead in both systems of representation (recall that I stressed the revisability of bridgeheads earlier). More precisely, it consisted in defining a new bridgehead leading from two-place representations of complex algebra to not three- but four-place systems — the systems that Hamilton quickly named *quaternions*. Thus, within the algebraic representation, the basic entities were extended from 2 to 4, from 1, i to 1, i, j, k, while within the geometrical representation, as Hamilton wrote the next day, "there dawned on me the notion that we must admit, in some sense, a *fourth dimension* of space" — with the fourth dimension, of course, mapped by the new k-axis.[38]

We will consider this shift in bridgehead further in the next section; for now, we can observe that Hamilton had still not completed his initial development of quaternions. The quantity k^2 remained undefined at this stage, as did the various products of i and j with k, except for those intrinsic to his new bridgehead $ij = k$. Hamilton fixed the latter products by a combination of filling assumptions and forced moves following from relations already fixed:

> I saw that we had probably $ik = -j$, because $ik = iij$ and $i^2 = -1$; and that in like manner we might expect to find $kj = ijj = -i$; from which I thought it likely that $ki = j, jk = i$, because it seemed likely

that if $ji = -ij$, we should have also $kj = -jk$, $ik = -ki$. And since the order of these imaginaries is not indifferent, we cannot infer that $k^2 = ijij$ is $+1$, because $i^2 \times j^2 = -1 \times -1 = +1$. It is more likely that $k^2 = ijij = -iijj = -1$. And in fact this last assumption is necessary, if we would conform the multiplication to the law of multiplication of moduli.[39]

Hamilton then checked whether the algebraic version of quaternion multiplication under the assumptions above, including $k^2 = -1$, led to results in accordance with the rule of multiplication concerning products of lengths in the geometrical representation ("the law of multiplication of moduli") and found that it did. Everything in his quaternion system was thus now defined in such a way that the laws of multiplication in both the algebraic and the geometrical version ran without resistance into one another. Through the move to four-place systems, Hamilton had finally found a successful accommodation to the resistances that stood in the way of his three-place extensions. The outcome of this dialectic was the general rule for quaternion multiplication:[40]

$$(a, b, c, d)(a', b', c', d') = (a'', b'', c'', d''),$$

where

$$a'' = aa' - bb' - cc' - dd',$$
$$b'' = ab' + ba' + cd' - dc',$$
$$c'' = ac' + ca' + db' - bd',$$
$$d'' = ad' + da' + bc' - cb'.$$

With these algebraic equations, and the geometrical representation of them, Hamilton had, in a sense, achieved his goal of associating calculation with geometry. He had found vectors of extension of algebra and geometry that interactively stabilized one another, as I say, preserving in four dimensions the one-to-one association of elements and operations previously established in two dimensions. I could therefore end my narrative here. But before doing so, I want to emphasize that the qualifier "in a sense" is significant. It marks the fact that what Hamilton had achieved was a *local* association of calculation with geometry rather than a global one. He had constructed a one-to-one correspondence between a particular algebraic system and a particular geometric system, not an all-purpose link between algebra and geometry, considered as abstract, all-encompassing entities. And this remark makes clear the fact that one important aspect of Hamilton's achievement was to redefine, partially at least, the cultural space of future mathematical and scientific practice: more new associations remained to be made if quaternions were ever to be "delocalized" and linked into the overall flow of mathematical and scientific practice, requiring work that would, importantly, have been inconceivable prior to Hamilton's construction of quaternions.

As it happens, from 1843 on, Hamilton devoted most of his productive energies to this task, and both quaternions and the principle of noncommutation that they enshrined were progressively taken up by many sectors of the scientific and mathematical communities.[41] Here, I will discuss one last aspect of Hamilton's practice that can serve to highlight the locality of the association embodied in quaternions. Earlier, I described Hamilton's organizing aim as that of connecting calculation with geometry. And, in fact, quaternions did serve to bring algebraic calculation to a geometry — to the peculiar four-dimensional space mapped by 1, i, j, and k. Unfortunately, this was not the geometry for which calculation was desired. The promise of triplet — not quaternion — systems had been that they would bring algebra to bear upon the real, three-dimensional world of interest to mathematicians and physicists. In threading his way through the dialectic of resistance and accommodation, Hamilton had, in effect, left that world behind. Or, to put it another way, his practice, as so far described, had served to displace resistance rather than fully accommodating to it. Technical resistances in the development of three-place mathematical systems had been transmuted into a resistance in moving between Hamilton's four-dimensional world and the three-dimensional world of interest. It was not evident how the two worlds might be related to one another. This was one of the first problems that Hamilton addressed once he had arrived at his algebraic formulation of quaternions.

In his letter to John Graves, of 17 October 1843, Hamilton outlined a new geometrical interpretation of quaternions that served to connect them back to the world of three dimensions. This new interpretation was a straightforward but consequential redescription of the earlier four-dimensional representation. Hamilton's idea was to think of an arbitrary quaternion (a, b, c, d) as the sum of two parts: a real part, a, which was a pure real number and had no geometrical representation, and an imaginary part, the triplet $ib + jc + kd$, which was to be represented geometrically as a line-segment in three-dimensional space.[42] Having made this split, Hamilton was then in a position to spell out rules for the multiplication of the latter line-segments, which he summarized as follows:

Finally, we may always decompose the latter problem [the multiplication of two arbitrary triplets] into these two others; to multiply
two pure imaginaries which agree in direction, and to multiply
two which are at right angles with each other. In the first case, the
product is a pure negative, equal to the products of the lengths or
moduli with its sign changed. In the second case, the product is a
pure imaginary of which the length is the product of the lengths
of the factors, and which is perpendicular to both of them. The
distinction between one such perpendicular and its opposite may
be made by the rule of rotation [stated earlier in this letter].

There seems to me to be something analogous to *polarized intensity* in the pure imaginary part; and to *unpolarized energy* (indifferent to direction) in the real part of a quaternion: and thus we
have some slight glimpse of a future Calculus of Polarities. This is
certainly very vague, but I hope that most of what I have said above
is clear and mathematical.[43]

These strange rules for the multiplication of three-dimensional line-
segments — in which the product of two lines might be, depending
upon their relative orientation, a number or another line or some combination of the two — served to align quaternions with mathematical
and scientific practice concerning the three-dimensional world.[44] Nevertheless, the association of algebra with geometry remained local. No
contemporary physical theories, for example, spoke of entities in three-
dimensional space obeying Hamilton's rules. It therefore still remained
to find out in practice whether quaternions could be delocalized to the
point where they might become useful. With hindsight, one can pick
out from the rules of multiplication a foreshadowing of modern vector
analysis, with its "dot" and "cross" products, and in the references to
"polarized intensity" and "unpolarized energy" one can find a gesture
toward electromagnetic theory, where quaternions and vector analysis
found their first important use. But, as Hamilton wrote, unlike the
mathematics of quaternions, this "slight glimpse of [the] future" was, in
1843, "certainly very vague." It was only in the 1880s, after Hamilton's
death, that Josiah Willard Gibbs and Oliver Heaviside laid out the fundamentals of vector analysis, dismembering the quaternion system into
more useful parts in the process.[45] This key moment in the delocaliza-
tion of quaternions was also the moment of their disintegration.

I have come to the end of my story of Hamilton and quaternions,
and the analysis that I have interwoven with this narrative is complex

enough, I think, to warrant a general summary and even a little further elaboration.

My overall objective here has been to get to grips with the specifically conceptual aspects of scientific practice (I continue to use "scientific" as an umbrella term that covers mathematics). My point of departure has been the traditional one, namely, an understanding of conceptual extension as a process of modeling. Thus I have tried to show that complex algebra and its geometrical representation in the complex plane were both constitutive models in Hamilton's practice. But I have gone beyond the tradition in two ways. First, instead of treating modeling (metaphor, analogy) as a primitive term, I have suggested that it bears further analysis and decomposition into the three phases of bridging, transcription, and filling. I have exemplified these phases and their interrelation in Hamilton's work, and I have tried to show how the openness of modeling is tentatively cut down by human discretionary choices — by human agency, traditionally conceived — in bridging and filling, and by disciplinary agency — disciplined, machinelike human agency — in transcription. I have also exemplified the fact that these two aspects of modeling — active and passive from the perspective of the human actor — are inextricably intertwined inasmuch as the object of constructing a bridgehead, for example, is to load onto it disciplined practices already established around the base model. Conceptual practice thus has the quality of a dance of agency, in this case between the discretionary human agent and what I have been calling disciplinary agency. The constitutive part played by disciplinary agency in this dance guarantees that the free moves of human agents — bridging and filling — carry those agents along trajectories that cannot be foreseen, but that have to be found out in practice.

My second step beyond traditional conceptions of modeling has been to note that it does not proceed in a vacuum. Issues of cultural multiplicity surface here. My suggestion is that conceptual practice is organized around the production of associations, the making (and breaking) of connections and the creation of alignments between disparate cultural elements, such as, in the present instance, the association (initially, at least) between three-place algebras and three-dimensional geometries. And the key observation is that the entanglement of disciplinary agency in conceptual practice makes the achievement of such associations nontrivial in the extreme. Hamilton wanted to extend algebra and geometry into three dimensions while maintaining a one-to-one correspondence between their respective elements and operations, but neither he nor anyone subsequently has been able to do so. Resistance thus emerges in conceptual practice in relation to intended associations and precipitates the dialectics of accommodation and further resistance that I call the mangle. Now I want to discuss just what gets mangled.

Most obviously mangled in Hamilton's practice were the modeling vectors that he pursued. In the face of resistance, he tinkered with choices of bridgeheads and fillings, tuning, one can say, the directions along which complex algebra and its geometrical representation were to be extended. And, as we saw, this mangling of modeling vectors eventually (not at all necessarily) met with success. In the quaternion system, Hamilton arrived at an association of one-to-one correspondence between the respective elements and operations of an extension of complex algebra and an extension of its geometrical representation. This achievement constituted an *interactive stabilization* of the specific free moves and the related forced moves that led up to it. *This* particular bridgehead coupled with *these* particular transcriptions and fillings defined the vector along which complex algebra should be extended, and similarly for the associated geometry. Exactly how existing conceptual structures should be extended was, then, the upshot of the mangle — as was the precise structure of the quaternion system that these particular extensions defined.

Here it is worth pausing to reiterate for conceptual practice two points that I have made elsewhere with respect to captures and framings of material agency. First, the precise trajectory and endpoint of Hamilton's practice were in no way given in advance. Nothing prior to that practice determined its course. In the real time of his mathematical work, Hamilton had to fix bridgeheads and fillings and to find out where they led via disciplined transcriptions. Furthermore, he had to find out, also in real time, just what resistances would emerge relative to his intended conceptual alignments — resistances that again could not be foreseen — and to make whatever accommodations he could to them, with the success or failure of such accommodations itself becoming apparent only in practice. Conceptual practice, therefore, has to be seen as *temporally emergent,* as do its products.[46] Likewise, it is appropriate to note the *posthumanist* aspect of conceptual practice as exemplified in Hamilton's work. My analysis here, as elsewhere, entails a decentering of the human subject, although this time by a shift toward disciplinary agency rather than toward the material agency that has been at issue in other discussions.[47] It is not, of course, the case that Hamilton as a human agent disappears from my analysis. I have not sought to reduce him to an "effect" of disciplinary agency, and I do not think that that can sensibly be done. Hamilton's free moves were just as constitutive of his practice and its products as his forced ones were. It is rather that the center of gravity of my account is positioned *between* Hamilton as a classical human agent, a locus of free moves, and the disciplines that carried him along. To be more precise, at the center of my account is a dance of intertwined human and disciplinary agency, the steps of which traced out the trajectory of Hamilton's practice.

So far I have been talking about the transformation of modeling vectors and formalisms in conceptual practice. But more was mangled and interactively stabilized in our example than that, and I want to consider the intentional structure of Hamilton's work before returning to its disciplinary aspects. One must, I think, take seriously Hamilton's stated intention "to connect . . . *calculation* with *geometry*, through some *extension*, to *space of three dimensions*." One cannot otherwise make sense of the dialectics of resistance and accommodation that steered his practice through the open-ended space of modeling and eventually terminated in the quaternion system. What I want to stress, however, is that we should think of specific goals and purposes as situated *in the plane* of scientific practice. They are not entities that control practice from without. Thus Hamilton's goal was conceivable only within the cultural space where an association between complex algebra and geometry had already been constructed, and it was further transformed (i.e., reoriented toward four instead of three dimensions) in the real time of his practice, as part and parcel of the dialectics of resistance and accommodation that we have examined. Hamilton aimed at an association in three dimensions, but he finally achieved one in four via the shift in bridgehead implicit in the introduction of the new square root of -1 that he called k. Like the technical elements of scientific culture, then, goals themselves are always liable to mangling in practice.

From the intentional structure of human agency we can turn to its disciplined, repetitive, machinelike aspects. I have emphasized that Hamilton was carried along in his practice by disciplinary agency, and it was crucial to my analysis that his transcriptions were not discretionary. Such a lack of discretion is the precondition for the emergence of dialectics of resistance and accommodation. But it is also worth emphasizing that Hamilton evidently did exercise discretion in choosing the particular disciplines to which he would submit himself. Thus, throughout his practice, he maintained the first part of the geometrical rule already established for the multiplication of lines in the complex plane (that the length of the product of two line-segments was the product of their individual lengths). But it was crucial to his path to quaternions that at a certain point he simply abandoned the second part of the multiplication rule concerning the orientation of product lines in space. He did not attempt to transcribe this when thinking about the multiplication of two arbitrary triplets. Part of Hamilton's strategy of accommodation to resistance was, then, a selective and tentative modification of discipline — in this case, an *eliminative* one. Hamilton bound himself to part but not all of established routine practice.[48]

One can also understand Hamilton's introduction of noncommuting quantities into his extension of complex algebra as a selective modification of discipline, but in this case an *additive* one. He continued to follow standard practice as far as ordinary numbers were concerned — treating their products as indifferent to the order of terms to be multiplied — but he invented a quite new, nonroutine rule for the multiplication of his various square roots of -1. In such ways Hamilton both drew upon established routines to carry himself along and, as part of his accommodation to resistance, transformed those routines, eliminating or adding to them as seemed to him promising. Disciplinary agency, I therefore want to say, has again to be seen as in the plane of practice and mangled there in the very dialectics of resistance and accommodation to which it gives structure. And, further, transformed disciplines are themselves interactively stabilized in the achievement of cultural associations. That certain specific transformations of discipline rather than others should have been adopted was itself determined by the specific association of calculation with geometry that Hamilton eventually achieved with quaternions.

I opened this essay by suggesting that one cannot claim to understand scientific practice unless one can offer an analysis of its specifically conceptual aspects, and that is what I have just sketched out. Now I want to indicate briefly how this analysis contributes to a more general picture of scientific practice and culture.

This essay is part of a project which seeks to develop an understanding of science in a performative idiom, an idiom capable of recognizing that the world is continually doing things and that so are we (in contrast to the traditional representational image of science as being, above all, in the business of representing a dead nature devoid of agency). Thus I have elsewhere paid close attention to the machines and instruments that are integral to scientific culture and practice, and I have concluded that we should see the machinic field of science as being precisely adjusted in its material contours to capture and frame material agency. The exact configuration of a machine or an instrument is the upshot of a tuning process that delicately positions it within the flow of material agency, harnessing and directing that agency — domesticating it. The image that lurks in my mind seems to be that of a finely engineered valve that both regulates and directs the flow of water from a pipe (though perhaps it is some kind of a turbine). The performative idiom also encourages us to think about human agency, and the argument that I have sought to exemplify here and elsewhere is that this can be grasped along similar lines. One should think of the scale and social relations of scientific agency, and the disciplined practices of such agency, as likewise being finely tuned in relation to its performativity; and, beyond that, of the engineering of the material and the human as not proceeding independently of one another: in scientific culture particular configurations of material and human agency appear interactively stabilized against one another.[49]

Once one begins to think about knowledge as well as performance, the picture becomes more elaborate, but its form remains the same. One can think of factual and theoretical knowledge in terms of representational chains passing through various levels of abstraction and conceptual multiplicity and terminating, in the world, on captures and framings of material agency. And, as we have seen here, conceptual structures (scientific theories and models, mathematical formalisms) can themselves be understood as positioned in fields of disciplinary agency much as machines are positioned in fields of material agency. Conceptual structures are like precisely engineered valves, too, domesticating disciplinary agency. Again, though, conceptual engineering should not be thought of as proceeding independently of the engineering of human and material agency. As I have just argued, disciplines, for instance, are themselves subject to transformation in conceptual practice, and, in general, the conceptual and material elements of culture should be seen as evolving together in empirical practice. Scientific culture, then, itself appears as a wild kind of machine built from radically heterogeneous parts, a supercyborg, harnessing material and disciplinary agency in material and human performances, some of which lead out into the world of representation, of facts and theories.[50]

I confess that I like this image of scientific culture. It helps me to fix in my mind the fact that the specific contents of scientific knowledge are always immediately tied to specific and precisely formed fields of machines and disciplines. Above all, it helps me to focus on the fact that scientific knowledge is just one part of the picture, something that is not analytically privileged in any way, but that evolves in an impure, posthuman dynamics together with all of the other cultural strata of science — material, human, social (in the next section I throw metaphysical systems into this assemblage, too). This understanding, of course, contrasts with traditional representationalist images of science, which can hardly get the nonrepresentational strata of science into focus and which can never grasp its performative aspect at all.

I turn now to the question of how the supercyborg of scientific culture is extended in time. Traditional answers assert that something substantive within scientific culture (as I define it) endures through cultural extension and explains or controls it—social interests, epistemic rules, or whatever. Or perhaps something quite outside culture has the controlling role: the world itself, Nature. Elsewhere, I have criticized the idea that the social can play the required explanatory role (an argument that I will take up below); I have also argued against any necessarily controlling role for epistemic rules given my own account of how "the world itself" plays into cultural extension.[51] Here I want to stress that on my analysis *nothing* substantive explains or controls the extension of scientific culture. Existing culture is the surface of emergence of its own extension, in a process of open-ended modeling having no destination that is given or knowable in advance. Everything within the multiple and heterogeneous culture of science is, in principle, at stake in practice. Trajectories of cultural transformation are determined in dialectics of resistance and accommodation played out in real-time encounters with temporally emergent agency, dialectics which occasionally arrive at temporary oases of rest in the achievement of captures and framings of agency and associations among multiple cultural extensions. I have noted, it is true, that one needs to think about the intentional structure of human agency to understand this process; vectors of cultural extension are tentatively fixed in the formulation of scientific plans and goals, and resistances have to be seen as relative to such goals. But, as I have shown, plans and goals are both emergent from existing culture and at stake in scientific practice, themselves liable to mangling in dialectics of resistance and accommodation. They do not endure through, explain, or control cultural extension.

So this is my overall claim about science: there is no substantive explanation to be given for the extension of scientific culture. There is, however, I would also claim, a temporal pattern to practice that we can grasp, that we can find instantiated everywhere, and that, once found and grasped, constitutes an understanding of what is going on. It is the pattern just described—of open-ended extension through modeling, dialectics of resistance and accommodation, and so on. And, in good conscience, this pattern—the mangle—is the only explanation that I can defend of what scientific culture becomes at any given moment, of the configuration of its machines, of its facts and theories, of its conceptual structures, disciplines, social relations, and so forth. Science mangles on.

The central task of this essay has been to understand how dialectics of resistance and accommodation can arise in conceptual practice. I want to end, however, by developing two subsidiary topics. It is common knowledge (among historians of mathematics, at least) that Hamilton was as much a metaphysician as a mathematician and that he felt that his metaphysics was, indeed, at the heart of his mathematics.[52] I therefore want to see how the relation between mathematics and metaphysics can be understood in this instance. At the same time, it happens that David Bloor has offered a clear and interesting explanation of Hamilton's metaphysics as a case study in the sociology of scientific knowledge (SSK). Since SSK has been at the heart of developments in science studies over the past thirty years or so, I think it will be useful to try to clarify how my own analysis of practice diverges from it.[53] In what follows I focus upon issues of temporal emergence and the possibilities for a distinctively sociological explanation of science (since this connects directly to the posthumanism of the mangle), beginning with Bloor's account of Hamilton's metaphysics and then offering my own.

In "Hamilton and Peacock on the Essence of Algebra," Bloor focuses upon the different metaphysical understandings of algebra articulated by Hamilton, on the one side, and by a group of Cambridge mathematicians, including Peacock, on the other. We can get at this difference by returning to the foundational crisis in nineteenth-century algebra. While the geometrical representation of complex algebra was one way of defusing the crisis and giving meaning to negative and imaginary quantities, some mathematicians did not opt for this commonsense route, preferring more metaphysical approaches. Peacock and his Cambridge colleagues took a *formalist* line, as Bloor calls it, which suggested that mathematical symbols and the systems in which they were embedded were sufficient unto themselves, in need of no extramathematical foundations and subject to whatever interpretation proved appropriate to specific uses. Thus, from the formalist point of view, there was and could be no foundational crisis in algebra. Nineteenth-century mathematician George Boole summed up this position as follows:

They who are acquainted with the present state of the theory of Symbolical Algebra are aware, that the validity of the processes of analysis does not depend upon the interpretation of the symbols which are employed, but solely upon the laws of their combination. Every system of interpretation which does not affect the truth of the relations supposed, is equally admissible, and it is thus that the same process may, under one scheme of interpretation, represent the solution of a question on the properties of numbers, under another, that of a geometrical problem, and under a third, that of a problem of dynamics or optics.[54]

Hamilton disagreed. He thought that mathematical symbols and operations must have some solid foundations that the mind latches onto — consciously or not — in doing algebra. And, as Bloor puts it:

> Hamilton's metaphysical interests placed him securely in the Idealist tradition. He adopted the Kantian view that mathematics is synthetic *a priori* knowledge. Mathematics derives from those features of the mind which are innate and which determine *a priori* the general form that our experience must take. Thus geometry unfolds for us the pure form of our intuition of space. Hamilton then said that if geometry was the science of pure space, then algebra was the science of pure time.[55]

Indeed, Hamilton developed his entire theory of complex algebra explicitly in such terms. In his "Theory of Conjugate Functions, or Algebraic Couples; With a Preliminary and Elementary Essay on Algebra as the Science of Pure Time," first read to the Royal Irish Academy in 1833, he showed how positive, real algebraic variables — denoted a, b, c, etc. — could be regarded as "steps" in time (rather than magnitudes of material entities) and how negative signs preceding them could be taken as denoting reversals of temporality, changing before into after. Hamilton also elaborated the system of couples mentioned earlier. Written (a, b), these couples transformed as the usual complex variables did under the standard mathematical operations, but, importantly, the problematic symbol "i" was absent from them. Hamilton's claim was thus to have positively located and described the foundations of complex algebra in our intuitions of time and its passing.[56]

This much is well-known, but Bloor takes the argument one step further: "I am interested in why men who were leaders in their field, and who agreed about so much at the level of technical detail, nevertheless failed to agree for many years about the fundamental nature of their science. I shall propose and defend a sociological theory about Hamilton's metaphysics and the divergence of opinion about symbolical algebra to which he was a party."[57] Bloor's idea is thus, first, that the

technical substance of algebra did not determine its metaphysical interpretation and therefore, second, that we need to invoke something other
than technical substance — namely, the social — to explain why particular individuals and groups subscribed to the metaphysical positions that
they did. This is a standard opening gambit in SSK, and Bloor follows it
up by discussing the different social positions and visions of Hamilton
and the Cambridge formalists and explaining how particular metaphysical views served to buttress them.[58] According to Bloor, Hamilton was
aligned with Coleridge and his circle and, more broadly, with "the interests served by Idealism" — conservative, holistic, reactionary interests
opposing the growing materialism, commercialism, and individualism
of the early nineteenth century and the consequent breakdown of the
traditional social order.[59] As Bloor explains it, Hamilton's idealism as-
similated mathematics to the Kantian category of Understanding, which
was in turn understood to be subordinate to the higher faculty of Reason. And, on the plane of human affairs, Reason was itself the province
not of mathematics, but of religion and the Church. Thus the "practical
import" of Hamilton's idealism "was to place mathematics as a profes-
sion in a relation of general subordination to the Church. Algebra, as
Hamilton viewed it, would always be a reminder of, and a support for, a
particular conception of the social order. It was symbolic of an 'organic'
social order of the kind which found its expression in Coleridge's work
on Church and State."[60]

So, Hamilton's social vision and aspirations structured his metaphysics. As far as the Cambridge group of mathematicians was concerned, the same pattern was repeated but from a different starting
point. In mathematics and beyond, they were both "reformers and radi-
cals" and "professionals" keen to assert their autonomy from traditional
sources of authority like the Church.[61] Their formalism and its opposition to the foundationalism of people like Hamilton, then, served
this end, defining mathematics as the special province of mathematicians. It was an antimetaphysics, one might say, which served to keep
metaphysicians and the Church out. As Bloor summarizes his analysis of the differences between the two parties over the foundations of
mathematics:

> Stated in its broadest terms, to be a formalist was to say: "we can
> take charge of ourselves." To reject formalism was to reject this
> message. These doctrines were, therefore, ways of rejecting or en-
> dorsing the established institutions of social control and spiritual
> guidance, and the established hierarchy of learned professions and

intellectual callings. Attitudes towards symbols were themselves symbolic, and the messages they carried were about the autonomy and dependence of the groups which adopted them.[62]

I have no quarrel with Bloor's arguments as rehearsed so far. I have no knowledge of the social locations and aspirations of the parties concerned that would give me cause to doubt the existence of the social/ metaphysical/mathematical correlations he outlines. But still, something peculiar happens toward the end of Bloor's essay. He concludes by stating, "I do not pretend that this account is without problems or complicating factors," and then lists them.[63] For the rest of this section we will be concerned with just one complicating factor.

"It is necessary," Bloor remarks, "to notice and account for the fact that Hamilton's opposition to Cambridge formalism seemed to decline with time. In a letter to Peacock dated Oct. 13, 1846, Hamilton declared that his view about the importance of symbolical science 'may have approximated gradually to yours.' Interestingly," Bloor continues, "Hamilton also noted some four years later 'how much the course of time has worn away my political eagerness.'"[64] The structure of these sentences is, I think, characteristic of what ssk, the sociology of scientific knowledge, looks like when brought to bear upon empirical studies. Note first that the shift in Hamilton's metaphysics is viewed as a "problem." It appears that way to Bloor because he wants to understand the social not just as a correlate of the metaphysical, but as a kind of cause.[65] The social is the solid, reliable foundation that holds specific metaphysical positions in place in an otherwise open-ended space. Any drifting of Hamilton's metaphysics threatens this understanding, and Bloor therefore tries to recoup this drift, by qualifying it as perhaps apparent ("seemed to decline") and then by associating it with a decline in Hamilton's "political eagerness." Perhaps Hamilton's social situation and views changed first and gave rise to Hamilton's concessions to formalism, seems to be Bloor's message (though the dates hardly look promising). If so, Bloor's causal arrow running from the social to the metaphysical would be secure.[66]

Before offering a different interpretation of Hamilton's metaphysical wandering, I want to comment further on Bloor's general position. Three points bear emphasis. First, although ssk tends to regard the social as a nonemergent cause of cultural change in science, it is clear that Bloor *does* recognize here that the social can itself change with time. This is precisely how he hopes to account for the problem of changing metaphysics. But, second, he offers no examination or analysis of how the social changes. The social, I would say, is treated as at most a *quasi-emergent* category, both in this essay and in the SSK canon in general.[67] The SSK gaze only ever catches a fixed image of the social in the act of structuring the development of the technical and metaphysical strata of science. In other words, SSK always seems to miss the movie in which the social is itself transformed.[68] Bloor's essay, then, exemplifies an important difference between SSK and the mangle, for I would argue that the social should in general be seen as in the plane of practice, both feeding into technical practice and being emergently mangled there, rather than as a fixed origin of unidirectional, causal arrows. Third, as is characteristic of SSK, Bloor does not even consider the possibility that there might be any explanation for Hamilton's metaphysical shift *other than* a change in the social. In contrast, I now want to offer an explanation that refers this shift not outward to the social, but inward, toward Hamilton's technical practice.

Bloor says that Hamilton's opposition to formalism "seemed" to decline, but the evidence is that there was no seeming about it. As Hamilton put it in another letter, written in 1846 to his friend Robert P. Graves,

> I feel an increased sympathy with, and fancy that I better understand, the Philological School [Bloor's formalists]. It enables me to see better the high functions of language, to trace more distinctly and more generally the influence of signs over thoughts, and to understand an answer which I hazarded some years ago to a question of yours, What did I suppose to be the *Science of Pure Kind?* namely, that I supposed it must be the *Science of Symbols.*[69]

The year 1846, in fact, seems to have been an important one in Hamilton's metaphysical biography. It was just around then that he began to indicate in various ways that his position had changed. One might suspect, therefore, that Hamilton's worries about metaphysical idealism had their origins in his technical practice around quaternions in the early 1840s.[70] And this suspicion is supported by the fact that Hamilton's technical writings on quaternions — specifically, the preface to his first book on the subject, the massive *Lectures on Quaternions* of 1853 — contain several explicit discussions of his past and present metaphysical stances. We can peruse a few of these and try to make sense of what happened.

Hamilton's preface to the *Lectures* took the form of a historical introduction to his thought and to related work by other mathematicians, and one striking feature of it is the tone of regret and retraction that

Hamilton adopted whenever the Science of Pure Time came up. The preface begins with a summary of his early work on couples, introduced with the following remark:

> In this manner I was led, many years ago, to regard Algebra as the SCIENCE OF PURE TIME. . . . If I now reproduce a few of the opinions put forward in that early Essay, it will be simply because they may assist the reader to place himself in that *point of view,* as regards the first elements of *algebra,* from which a passage was gradually made by me to that comparatively *geometrical* conception which it is the aim of this volume to unfold. And with respect to anything unusual in the *interpretations* thus proposed, for some simple and elementary notations, it is my wish to be understood as not at all insisting on them as *necessary,* but merely proposing them as consistent amongst themselves, and preparatory to the study of quaternions, in at least one aspect of the latter.[71]

So much for a priori knowledge. Later, Hamilton's tone verges upon apology for mentioning his old metaphysics: "Perhaps I ought to apologise for having thus ventured here to reproduce (although only historically . . .) a view so little supported by scientific authority. I am very willing to believe that (though not unused to calculation) I may have habitually attended too little to the *symbolical* character of Algebra, as a Language, or organized system of *signs:* and too much (in proportion) to what I have been accustomed to consider its *scientific* character, as a Doctrine analogous to Geometry, through the Kantian parallelism between the *intuitions* of Time and Space."[72] Later still, Hamilton speaks positively about the virtues of formalism and their integration into his own mathematical practice, saying that he "had attempted, in the composition of that particular series" (referring to papers on quaternions understood as quotients of lines in three-dimensional space, published between 1846 and 1849) "to allow a more prominent influence to the general *laws of symbolical language* than in some former papers of mine; and that to this extent I had on this occasion sought to imitate the *Symbolical Algebra* of Dr Peacock."[73]

Far from being situated on the opposite side of a metaphysical gulf from Peacock, then, by 1846 Hamilton was *imitating* Peacock's formalist approach in his technical practice (without, I should add, entirely abandoning his earlier Kantianism). And to understand why, we need to look more closely at that practice. In the very long footnote that begins with his apology for mentioning the Science of Pure Time, Hamilton actually goes on to assert that he could have developed many aspects of the quaternion system to be covered in the rest of the book within his original metaphysical framework and that this line of development "would offer no result which was not perfectly and easily *intelligible,* in strict consistency with that *original* thought (or intuition) of time, from which the whole theory should (on this supposition) be evolved. . . . Still," he continues,

> I admit fully that the actual *calculations* suggested by this [the Science of Pure Time], or any other view, must be performed according to some fixed *laws of combination of symbols,* such as Professor De Morgan has sought to reduce, for ordinary algebra, to the smallest possible compass . . . and that in following out such *laws* in their symbolical consequences, uninterpretable (or at least uninterpreted) *results* may be expected to arise. . . . [For example,] in the passage which I have made (in the Seventh Lecture), from *quaternions* considered as *real* (or as geometrically *interpreted*), to *biquaternions* considered as *imaginary* (or as geometrically *uninterpreted*), but as symbolically *suggested* by the generalization of the quaternion formulae, it will be perceived . . . that I have followed a *method of transition,* from *theorems proved* for the *particular* to *expressions assumed* for the *general,* which bears a very close *analogy* to the methods of Ohm and Peacock: although I have *since* thought of a way of *geometrically interpreting the biquaternions* also.[74]

Now, I am not going to exceed my competence by trying to explain what biquaternions are and how they specifically fit into the story, but I think one can get an inkling from this passage of how and why Hamilton's metaphysics changed. While Hamilton had found it possible to calmly work out his version of complex algebra on the basis of his Kantian notions about time, in his subsequent mathematical practice leading through quaternions he was, to put it crudely, flying by the seat of his pants. He was struggling through dialectics of resistance and accommodation, reacting as best he could to the exigencies of technical practice, without much regard to or help from any a priori intuitions of the inner meanings of the symbols he was manipulating. The variety of the bridging and filling moves that he made on the way to quaternions (reviewed above), for example, hardly betrays any "strict consistency" with an "*original* thought (or intuition)." Further, what guided Hamilton through the open-ended space of modeling was disciplinary agency — the replaying of established *formal* manipulations in new contexts marked out by bridging and filling. And, at the level of products rather than processes, a similar situation obtained. Hamilton continually arrived at technical results and then had to scratch around for interpretations of them — starting with the search for a three-dimensional geometric interpretation of his initial four-dimensional formulation of quaternions and ending up with biquaternions ("I have since thought of a way"). Moreover, Hamilton was in fact able to think of several ways of interpreting his findings. In the preface to the *Lectures* he discusses three different three-dimensional geometrical interpretations, one of which (not that mentioned in an earlier section here) formed the basis for his exposition of quaternions in the body of the book.[75] Formal results followed by an indefinite number of interpretations: this is a description of formalist metaphysics.

So, there is a prima facie case for understanding the transformation in Hamilton's metaphysics in the mid-1840s as an accommodation to resistances arising in technical-metaphysical practice. A tension emerged between Hamilton's Kantian a priorism and his technical practice, to which he responded by attenuating the former and adding to it an important dash of formalism. My suggestion is, therefore, that we should see metaphysics as yet another heterogeneous element of the culture in and on which scientists operate. Like the technical culture of science — and like the conceptual, like the social, like discipline — metaphysics is itself at stake in practice and just as liable to temporally emergent mangling there in interaction with all of those other elements. That is the positive conclusion of this section as far as my analysis of practice is concerned.

Comparatively, I have tried to show how my analysis differs from SSK in its handling of a specific example. Where SSK necessarily looks outward from metaphysics (and technical culture in general) to quasi-emergent aspects of the social for explanations of change (and stability), I have looked inward, to technical practice itself. There is an emergent dynamics in that practice which goes unrecognized by SSK. I have, of course, said nothing on my own account about the transformation in Hamilton's "political eagerness" that Bloor mentions. Having earlier argued for the mangling of the social, I find it quite conceivable that Hamilton's political views might also have been emergently mangled and interactively stabilized alongside his metaphysics in the evolution of the quaternion system. On the other hand, they might not. I have no more information on this topic than Bloor does — but at least the mangle can indicate a way to move past the peculiar quasi-emergent vision of the social that SSK offers us.

NOTES

1   Questions of agency in science have been thematized most clearly and insistently
    in the actor-network approach developed by Michel Callon, Bruno Latour, and John
    Law. See, for example, Callon and Latour, "Don't Throw the Baby Out with the Bath
    School! A Reply to Collins and Yearley," in *Science as Practice and Culture*, ed. Andrew
    Pickering (Chicago, 1992), 343–68. But Latour is right to complain about the dearth
    of studies and analyses of conceptual practice in science: "Almost no one," as he
    puts it, "has had the courage to do a careful anthropological study"; see his *Science
    in Action* (Cambridge, MA, 1987), 246. Whether failure of nerve is quite the prob-
    lem, I am less sure. Much of the emphasis on the material dimension of science
    in recent science studies must be, in part, a reaction against the theory-obsessed
    character of earlier history and philosophy of science. In any event, Eric Livingston's
    Ethnomethodological Foundations of Mathematics (Boston, 1986) is a counterexample
    to Latour's claim, and the analysis of conceptual practice that follows is a direct
    extension of my own earlier analysis of the centrality of modeling to theory develop-
    ment in elementary-particle physics. See Andrew Pickering, "The Role of Interests
    in High-Energy Physics: The Choice Between Charm and Colour," in *The Social Pro-
    cess of Scientific Investigation, Vol. 4 of Sociology of the Sciences*, ed. Karin Knorr, Roger
    Krohn, and Richard Whitley (Dordrecht, 1981), 107–38; and *Constructing Quarks: A
    Sociological History of Particle Physics* (Chicago, 1984). Nevertheless, resistance and
    accommodation were not thematized in my earlier analyses, and it may be that this
    exemplifies the lack of which Latour complains.
2   Andrew Pickering, *The Mangle of Practice: Time, Agency and Science* (Chicago, 1995),
    chapters 2, 3, and 5. This essay is a slightly revised version of chapter 4. I thank
    Barbara Herrnstein Smith for her editorial suggestions.
3   The Wittgenstein quotation (with his emphases) is taken from Michael Lynch, "Ex-
    tending Wittgenstein: The Pivotal Move from Epistemology to the Sociology of
    Science," in Pickering, ed., *Science as Practice*, 289. Lynch's commentary continues:
    "If the 'use' is the 'life' of an expression, it is not as though a meaning is 'attached' to
    an otherwise lifeless sign. We first encounter the sign in use or against the backdrop
    of a practice in which it has a use. It is already a meaningful part of the practice,
    even if the individual needs to learn the rule together with the other aspects of
    the practice. It is misleading to ask 'how we attach meaning' to the sign, since the
    question implies that each of us separately accomplishes what is already established
    by the sign's use in the language game. This way of setting up the problem is like
    violently wresting a cell from a living body and then inspecting the cell to see how
    life would have been attached to it."
4   Harry Collins, *Artificial Experts: Social Knowledge and Intelligent Machines* (Cam-
    bridge, MA, 1990).
5   The notion of discipline as a performative agent might seem odd to those accus-
    tomed to thinking of discipline as a constraint upon human agency, but I want (like
    Foucault) to recognize that discipline is productive. There could be no conceptual
    practice without the kind of discipline at issue; there would only be marks on paper.
6   See Pickering, "The Role of Interests," and *Constructing Quarks*; for the literature on
    metaphor and analogy in science more generally, see Barry Barnes, *T. S. Kuhn and
    Social Science* (London, 1982); David Bloor, *Knowledge and Social Imagery*, 2d ed. (Chi-
    cago, 1991 [1976]); Mary Hesse, *Models and Analogies in Science* (Notre Dame, 1966);

Karin Knorr-Cetina, *The Manufacture of Knowledge: An Essay on the Constructivist and Contextual Nature of Science* (Oxford, 1981); and Thomas Kuhn, *The Structure of Scientific Revolutions*, 2d ed. (Chicago, 1970 [1962]).

7   Ludwik Fleck, *Genesis and Development of a Scientific Fact* (Chicago, 1979).

8   Yves Gingras and S. S. Schweber, "Constraints on Construction," *Social Studies of Science* 16 (1986): 380.

9   A point of clarification in relation to my earlier writings might be useful. In "Living in the Material World: On Realism and Experimental Practice," in *The Uses of Experiment: Studies in the Natural Sciences*, ed. David Gooding, Trevor Pinch, and Simon Schaffer (Cambridge, 1989), 275–97, I discussed the open-ended extension of scientific culture in terms of a metaphor of "plasticity." I said that cultural elements were plastic resources for practice. The problem with this metaphor is that, if taken too seriously, it makes scientific practice sound too easy—one just keeps molding the bits of putty until they fit together. The thrust of my discussion of disciplinary agency here is that, unlike putty, pieces of conceptual culture keep transforming themselves in unpredictable ways after one has squeezed them. (Evidently, the same can be said of machine culture: one can tinker with the material configuration of an apparatus, but that does not determine how it will come to perform.) This is why achieving associations in practice is really difficult (and chancy).

10 Latour, *Science in Action*, 239, 241, and 242.

11 I should mention one important aspect of mathematics that distinguishes it from science and to which I cannot pay detailed attention here, namely, mathematical *proof*. Here, Imre Lakatos's account of the development of Euler's theorem points once more to the mangle in conceptual practice; see his *Proofs and Refutations: The Logic of Mathematical Discovery* (Cambridge, 1976). The exhibition of novel counterexamples to specific proofs of the theorem counts, in my terminology, as the emergence of resistances, and Lakatos nicely describes the revision of proof procedures as open-ended accommodation to such resistances, with interactive stabilization amounting to the reconciliation of such procedures to given counterexamples. Other work in the history and philosophy of mathematics that points toward an understanding of practice as the mutual adjustment of cultural elements includes Philip Kitcher, *The Nature of Mathematical Knowledge* (Oxford, 1983); and "Mathematical Naturalism," in *History and Philosophy of Modern Mathematics*, ed. William Aspray and Philip Kitcher (Minneapolis, 1988), 293–325, who argues that every mathematical practice has five components ("Mathematical Naturalism," 299); Michael Crowe, "Ten Misconceptions about Mathematics and Its History," in Aspray and Kitcher, eds., *Modern Mathematics*, 260–77; and "Duhem and History and Philosophy of Mathematics," *Synthese* 83 (1990): 431–47 (see note 29, below); and Gaston Bachelard, who understands conceptual practice in terms of "resistances" (his word) and "interferences" between disjoint domains of mathematics (see Mary Tiles, *Bachelard: Science and Objectivity* [Cambridge, 1984]).

12 I would have been entirely unaware of this were it not for the work of Adam Stephanides, then a graduate student nominally under my supervision. Stephanides brought Hamilton's work to my attention by writing a very insightful essay emphasizing the open-endedness of Hamilton's mathematical practice, an essay which eventually turned into Andrew Pickering and Adam Stephanides, "Constructing Quaternions: On the Analysis of Conceptual Practice," in Pickering, ed., *Science as Practice*, 139–67.

13 Sir William Rowan Hamilton, "Quaternions," Notebook 24.5, entry for 16 October 1843; and "Letter to Graves on Quaternions; or on a New System of Imaginaries in Algebra" (17 Oct. 1843), *Philosophical Magazine* 25 (1843): 489–95; both reprinted in Hamilton, *The Mathematical Papers of Sir William Rowan Hamilton*, Vol. 3, *Algebra* (Cambridge, 1967), 103–5, 106–10. (All page number citations of these and other writings by Hamilton refer to this 1967 edition of his papers.) My primary source of documentation is a first-person narrative written after the event, so it must be understood as an edited rather than a complete account (whatever the latter might mean). Nevertheless, it is sufficient to exemplify the operation of the mangle in conceptual practice, which is my central concern.

14 Thomas L. Hankins, *Sir William Rowan Hamilton* (Baltimore, 1980), 295.

15 Ibid., 295–300; J. O'Neill, "Formalism, Hamilton and Complex Numbers," *Studies in History and Philosophy of Science* 17 (1986): 351–72; Helena Pycior, "The Role of Sir William Rowan Hamilton in the Development of Modern British Algebra" (Ph.D. dissertation, Cornell University, 1976), chapter 7; B. L. van der Waerden, "Hamilton's Discovery of Quaternions," *Mathematics Magazine* 49 (1976): 227–34; and E. T. Whittaker, "The Sequence of Ideas in the Discovery of Quaternions," *Royal Irish Academy, Proceedings* 50, A6 (1945): 93–98.

16 See Hankins, *Sir William Rowan Hamilton*, 248; and Pycior, "The Role of Sir William Rowan Hamilton," chapter 4.

17 See Michael J. Crowe, *A History of Vector Analysis: The Evolution of the Idea of a Vectorial System* (New York, 1985), 5–11.

18 At this point my analysis must get technical, but readers whose familiarity with mathematics is limited could skim the rest of this section and the following one, returning to them if aspects of the subsequent discussion seem obscure. The important thing is to grasp the overall form of the analysis in these two sections rather than to follow all of Hamilton's mathematical maneuvers in detail.

19 The easiest way to grasp these rules is as follows. In algebraic notation, any complex number $z = x + iy$ can be reexpressed as $r(\cos\theta + i\sin\theta)$, which can in turn be reinterpreted geometrically as a line-segment of length r, subtending an angle θ with the x-axis at the origin. The product of two complex numbers z_1 and z_2 is therefore $r_1 r_2(\cos\theta_1 + i\sin\theta_1)(\cos\theta_2 + i\sin\theta_2)$. When the terms in brackets are multiplied out and rearranged using standard trigonometric relationships, one arrives at $z_1 z_2 = r_1 r_2[\cos(\theta_1 + \theta_2) + i\sin(\theta_1 + \theta_2)]$, which can itself be reinterpreted geometrically as a line-segment having a length that is the product of the lengths of the lines to be multiplied (part [a] of the rule) and making an angle with the x-axis equal to the sum of angles made by the lines to be multiplied (part [b]).

20 Sir William Rowan Hamilton, preface to *Lectures on Quaternions* (Dublin, 1853); reprinted in *Mathematical Papers*, 135; his emphases. The perceived need for an algebraic system that could represent elements and operations in three-dimensional space more perspicuously than existing systems is discussed in Crowe, *Vector Analysis*, 3–12. Although Hamilton wrote of his desire to connect calculation with geometry some years after the event, he recalled in the same passage that he was encouraged to persevere, despite his difficulties, by his friend John T. Graves, "who felt the wish, and formed the project, to surmount them in some way, as early, or perhaps earlier than myself" (*Mathematical Papers*, 137). Hamilton's common interest with Graves in algebra dated back to the late 1820s (see Hankins, *Sir William Rowan Hamilton*, chap. 17), so there is no reason to doubt that this utilitarian interest

did play a role in Hamilton's practice. See also O'Neill, "Formalism, Hamilton and Complex Numbers."

21   Hamilton, *Mathematical Papers,* 3–100, 117–42; Hankins, *Sir William Rowan Hamilton,* 245–301; Pycior, "The Role of Sir William Rowan Hamilton," chapters 3 to 6.

22   Hamilton, "Quaternions," 103; his emphases.

23   Hamilton, *Lectures on Quaternions,* 126–32. In such attempts, the intention to preserve any useful association of algebra and geometry does not seem to have been central: Hamilton's principal intent was simply to model the development of a three-place algebraic system on his existing two-place system of couples. Because the construction of associations in a multiple field plays a key role in my analysis, I should note that attention to this concept illuminates even these principally algebraic attempts. Hamilton found it necessary to transcribe parts of his development of couples *piecemeal,* and the goal of reassembling (associating) the disparate parts of the system that resulted again led to the emergence of resistance.

24   The foundational significance of Hamilton's couples was precisely that the symbol i did not appear in them and was therefore absent from the attempts at triplets discussed in note 23. A typical bridging move in those attempts was to go from couples written as (a, b) to triplets written as (a, b, c).

25   Hamilton, "Quaternions," 103; his emphases.

26   See, for example, his development of rules for the multiplication of couples: Sir William Rowan Hamilton, "Theory of Conjugate Functions, or Algebraic Couples; With a Preliminary and Elementary Essay on Algebra as the Science of Pure Time," *Transactions of the Royal Irish Academy* 17 (1837): 293–422; reprinted in *Mathematical Papers,* 3–96.

27   In *Lectures on Quaternions,* 139–40, Hamilton cites his notes of 1830 as containing an attempt at constructing a geometrical system of triplets by denoting the end of a line-segment in spherical polar coordinates as $x = r\cos\theta$, $y = r\sin\theta\cos\phi$, $z = r\sin\theta\sin\phi$, and extending the rule of multiplication from two to three dimensions as $r'' = rr'$, $\theta'' = \theta + \theta'$, $\phi'' = \phi + \phi'$. This addition rule for the angle ϕ breaks the coplanarity requirement at issue.

28   One route to this result is to write the triplet t in spherical polar notation. According to the rule just stated, on squaring, the length of the line-segment goes from r to r^2 and the angle θ doubles, while the angle ϕ remains the same. Using standard trigonometric relations to express $\cos 2\theta$ and $\sin 2\theta$ in terms of $\cos\theta$ and $\sin\theta$, one can then return to x, y, z notation and arrive at equation (2).

29   The Duhem problem is usually formulated in terms of open-ended responses to mismatches between scientific data and theoretical predictions; see Pierre Duhem, *The Aim and Structure of Physical Theory* (Princeton, 1991). As far as I am aware, the only prior discussion of it as it bears on purely mathematical/conceptual practice is to be found in Crowe, "Duhem," who argues, following Lakatos, *Proofs and Refutations,* that contradictions between proofs and counterexamples need not necessarily disable the former. In "Ten Misconceptions," Crowe also discusses the extension of Duhem's ideas about physics to mathematics and gestures repeatedly, although without detailed documentation or analysis, to the interactive stabilization of axioms and theorems proved within them, and of mathematical systems and the results to which they lead. I thank Professor Crowe for drawing my attention to these essays.

30   The following day, Hamilton described the idea of setting $ij = 0$ as "odd and uncomfortable" ("Letter to Graves," 107). He offered no reasons for this description, and it is

perhaps best understood as written from the perspective of his subsequent achievement. The quaternion system preserved the geometrical rule of multiplication that the length of the product was the product of the lengths of the lines multiplied. Since in the geometrical representation both i and j have unit length, the equation $ij = 0$ violates this rule. Here we have a possible example of the retrospective reconstruction of accounts in the rationalization of free moves.

31 In "The Role of Sir William Rowan Hamilton," 147, Pycior notes that Hamilton had been experimenting with noncommuting algebras as early as August 1842, although he then tried the relations $ij = j$, $ji = i$. Hankins, in *Sir William Rowan Hamilton*, 292, detects the possible influence of a meeting between Hamilton and the German mathematician Gotthold Eisenstein in the summer of 1843.

32 If one multiplies out the terms of equation (1), paying attention to the order of factors, the coefficient of yz in the last term on the right-hand side becomes $(ij + ji)$; Hamilton's assumption makes this coefficient zero.

33 Hamilton, "Quaternions," 103.

34 Ibid., 103–4.

35 According to Pythagoras's theorem, the square modulus of a line-segment is simply the sum of the squares of the coordinates of its endpoints, meaning the coefficients of 1, i, and j in algebraic notation.

36 Strictly speaking, this is too deterministic a formulation. The question really was whether any amount of tinkering with bridgeheads, fillings, and so on, could get past this point without calling up this or another resistance.

37 Hamilton, "Quaternions," 104; his emphasis.

38 Hamilton, "Letter to Graves," 108; his emphasis.

39 Ibid.

40 Ibid. Hamilton's notation for an arbitrary quaternion was (a, b, c, d). In the geometrical representation, the coordinates of the endpoint of a line-segment in four-dimensional space are given here; in algebraic notation this same quaternion would be written as $a + ib + jc + kd$.

41 See Hankins, *Sir William Rowan Hamilton*, chapter 23; and Crowe, *Vector Analysis*, chapters 4 to 7.

42 The origin of this distinction between real and imaginary parts of quaternions lay in the differences between the respective equations defining quaternion multiplication: the equation defining a'' has three minus signs, while those defining b'', c'', and d'' have two plus signs and only one minus.

43 Hamilton, "Letter to Graves," 110; his emphases.

44 Note that this geometric interpretation included a handedness rule—a "rule of rotation"—which reversed the sign of the product of perpendicular lines when the order of their multiplication was reversed, thus explaining algebraic noncommutation in much the same way as the two-dimensional geometrical representation of complex numbers had explained the "absurd" negative and imaginary quantities.

45 See Crowe, *Vector Analysis*, chapter 5.

46 Barbara Herrnstein Smith has commented to me on the preceding sentences that "idioms of *discovery* (things 'appearing,' agents 'finding' them) seem to dominate just where one might expect those of *construction* to emerge (for better or for worse) most emphatically." There is a point of potential confusion here that can be clarified. I do want to insist that scientists have to "find out," in real time, where practice leads, what resistances will emerge and relative to which associations and alignments. But

nothing in my analysis requires or supports the correspondence or Platonist realist assumption that a unique preexisting structure ("things") is exposed or discovered in the achievement of associations. I argue at some length against correspondence realism and in favor of a non-correspondence "pragmatic realism" (not antirealism) in *The Mangle of Practice*, chapter 6. In this connection, I can remark that when discussing Hamilton's unsuccessful attempts at constructing triplet systems, historians often invoke, in a Platonist fashion, later mathematical existence proofs that appear to be relevant. Thus, for example, Hankins (*Sir William Rowan Hamilton*, 438, n. 2) reproduces the following quotation from the introduction to volume 3 of Hamilton's collected papers (xvi): "Thirteen years after Hamilton's death G. Frobenius proved that there exist precisely three associative division algebras over the reals, namely, the real numbers themselves, the complex numbers and the real quaternions." One is tempted to conclude from such assertions that Hamilton's search for triplets was doomed in advance (or fated to arrive at quaternions) and that the temporal emergence of his practice and its products is therefore only apparent. Against this, one can note that proofs like Frobenius's are themselves the products of sequences of practices which remain to be examined. There is no reason to expect that analysis of these sequences would not point to the temporal emergence of the proofs themselves. Note also that these sequences were precipitated by Hamilton's practice and by subsequent work on triplets, quaternions, and other many-place systems, all of which served to mark out what an "associative division algebra over the reals" might mean. Since this concept was not available to Hamilton, he cannot have been looking for new instances of it. On the defeasibility of "proof," see Lakatos, *Proofs and Refutations;* and Trevor Pinch, "What Does a Proof Do If It Does Not Prove? A Study of the Social Conditions and Metaphysical Divisions Leading to David Bohm and John von Neumann Failing to Communicate in Quantum Physics," in *The Social Production of Scientific Knowledge*, Vol. 1 of *Sociology of the Sciences*, ed. Everett Mendelsohn, Peter Weingart, and Richard Whitley (Dordrecht, 1977), 171–215.

47   See Pickering, *The Mangle of Practice*, chapters 2, 3, 5.

48   Similarly, when Hamilton encountered resistances in his earlier attempts to construct an algebraic system of triplets modeled on his system of couples, he abandoned the established algebraic principle of unique division; see Hamilton, *Mathematical Papers*, 129–31.

49   See Pickering, *The Mangle of Practice*, chapters 2, 3.

50   I have seen draft essays by John Law that evoke a related image of scientific and technological culture as a kind of giant plumbing system, continually undergoing extension and repair.

51   See Pickering, *The Mangle of Practice*, section 2.5 and chapter 6, respectively.

52   See Hankins, *Sir William Rowan Hamilton;* and John Hendry, "The Evolution of William Rowan Hamilton's View of Algebra as the Science of Pure Time," *Studies in History and Philosophy of Science* 15 (1984): 63–81.

53   David Bloor, "Hamilton and Peacock on the Essence of Algebra," in *Social History of Nineteenth Century Mathematics*, ed. H. Mehrtens, H. Bos, and I. Schneider (Boston, 1981), 202–32. Canonical works in SSK include Barnes, *T. S. Kuhn;* Bloor, *Knowledge and Social Imagery;* Harry Collins, *Changing Order: Replication and Induction in Scientific Practice*, 2d ed. (Chicago, 1992 [1985]); and Steven Shapin, "History of Science and Its Sociological Reconstructions," *History of Science* 20 (1982): 157–211.

54   George Boole, *Mathematical Analysis of Logic* (1847); quoted in Ernest Nagel, *Tele-*

ology Revisited and Other Essays in the Philosophy and History of Science (New York, 1979), 166.

55 Bloor, "Hamilton and Peacock," 204.

56 Hendry, in "The Evolution of William Rowan Hamilton's View of Algebra," offers a subtle analysis of the early development of Hamilton's Kantianism and suggests that Hamilton might possibly have developed his theory of couples independently of it, hitching the metaphysics to the algebra "as a vehicle through which to get the essay published" (64). This observation creates no problems for my argument in what follows. I do not insist that practice always has a metaphysical flavor, but I do want to insist that metaphysics, when relevant, is subject to change and transformation in practice.

57 Bloor, "Hamilton and Peacock," 203.

58 The only respect in which Bloor's essay is untypical of SSK is his stopping short, that is, explaining metaphysics without pressing on to the technical substance of science. He does remark, however, that "should it transpire that this metaphysics is indeed relevant to technical mathematics, then my ideas may help to illuminate these matters as well" (ibid., 206). I am more concerned with the overall form of Bloor's argument than with its restriction to metaphysics.

59 Ibid., 220.

60 Ibid., 217.

61 Ibid., 222, 228.

62 Ibid., 228.

63 Ibid.

64 Ibid., 229.

65 In *Knowledge and Social Imagery*, 7, Bloor lists causal social explanation as the first distinguishing mark of the "strong programme" in SSK.

66 Thus Bloor's text, immediately following this quotation, continues: "A corresponding and opposite movement took place in Whewell's life. Here, *in obliging conformity with my thesis*, it is known that as Whewell moved to the right . . . he increasingly moved away from the symbolical approach in his mathematical writings" ("Hamilton and Peacock," 229–30; my emphases).

67 See the works cited in note 53.

68 In its early development, SSK was articulated against philosophical positions that rancorously opposed the suggestion that there was *anything* significantly social about scientific knowledge. A concentration on situations where the social could plausibly be regarded both as fixed and as explanatory of metaphysical and technical developments, therefore, fulfilled a strategic argumentative function for SSK. Its endless deferral of any inquiry into how the social might itself evolve seems strange, though, even given that background. It is, I suspect, part and parcel of SSK's almost principled refusal to interrogate key sociological concepts like "interest." Thus, almost two decades ago, Barry Barnes was writing in *Interests and the Growth of Knowledge* (London, 1977) that "new forms of activity arise not because men are determined by new ideas, but because they actively deploy their knowledge in a new context, as a resource to further their interests" (78), but then, on the last page of the book, shuffling interests out into unexplored regions of social theory with the remark, "I have deliberately refrained from advancing any precise definitions of 'interest' and 'social structure'; this would have had the effect of linking the claims being advanced to particular schools of thought within sociological theory. Instead, I have

been content, as it were, to latch the sociology of knowledge into the ongoing general trends of social thought" (86). Nothing has changed in the intervening years. Interests, and de facto their dynamics, are still left out in the cold by SSK. In a recent essay reviewing Latour's *Science in Action*, Steven Shapin writes: "One must . . . welcome any pressure that urges analysts further to refine, define, justify and reflect upon their explanatory resources. If there is misunderstanding, by no means all the blame needs to be laid at Latour's door. 'Interest-explanation' does indeed merit further justification"; see his "Following Scientists Around," *Social Studies of Science* 18 (1988): 549. And, replying to his critics in the "Afterword" to the second edition of *Knowledge and Social Imagery*, Bloor writes that "undeniably the terminology of interest explanations is intuitive, and much about them awaits clarification" (171). In chapter 3 of that same book, Bloor advances the Durkheimian argument that resistance to the strong programme arises from a sacred quality attributed to science in modern society. Perhaps in SSK the social has become the sacred.

69 Quoted in Nagel, *Teleology Revisited*, 189; Hamilton's emphases.

70 Hankins, in *Sir William Rowan Hamilton*, briefly connects Hamilton's metaphysical shift with his technical practice along the lines elaborated below (310). I should mention that Hamilton's Kantianism had a second string besides his thinking about time—namely, a concern with triadic structures grasped in relation to the Trinity (see 285–91). Besides possible utility, then, Hamilton's searches for triplets and his concern with three-dimensional geometry had themselves a metaphysical aspect. My focus here, though, is with his overall move away from Kantianism and toward formalism.

71 Hamilton, *Lectures on Quaternions*, 117–18; his emphases.

72 Ibid., 125; his emphases.

73 Ibid., 153; his emphases.

74 Ibid., 125–26; his emphases.

75 Ibid., 145–54. Of one of these systems, Hamilton wrote: "It seemed (and still seems) to me natural to connect this *extra-spatial unit* [the nongeometrical part of the quaternion] with the conception of TIME." But then he reverted to the formalist mode: "Whatever may be thought of these abstract and semi-metaphysical *views*, the *formulae* . . . are in any event a sufficient *basis* for the erection of a CALCULUS of quaternions" (152; his emphases).

The Moment of Truth on Dublin Bridge:

A Response to Andrew Pickering

Owen Flanagan

ANDREW Pickering's work is consistently interesting and important. His "Concepts and the Mangle of Practice: Constructing Quaternions" is no exception. Indeed, Pickering bravely and ably extends his own views about the social pragmatics of theory construction in science to mathematics to see if his model works in that more purely conceptual domain as it does in physics. Despite the fact that I find Pickering's general line persuasive, there are some issues to be joined. The three issues that I see as most critical to his project are (1) the question of what sort of analytic or explanatory device the *mangle of practice* is; (2) the question of *agency* in conceptual practice; and (3) the *metaphysical impulse* in mathematical thinking.

The Mangle. The mangle of practice is a metaphor. In "Objectivity and the Mangle of Practice," Pickering writes: "This dialectic of resistance and accommodation is *the mangle of practice* — my name for a genuinely emergent process that gives structure to the extension of scientific culture in the actual practice of scientific research. . . . The mangle is, I think, the single most important discovery made in the study of scientific practice and the consequences of its existence are dramatic and far-reaching. *It mangles just about everything in sight.*"[1]

I'm not sure that the mangle is an "it." There is no one kind of thing that a mangle is, so I'm uncomfortable with saying that "it" is "the genuinely emergent process that gives structure to the extension of scientific culture in the actual practice of scientific research." The basic insight is helpful. But I would say, *everything in sight gets mangled in the dialectic of resistance and accommodation,* instead of, "It [the mangle] mangles just about everything in sight." I prefer my way of expressing Pickering's point because it seems to capture the basic insight that every aspect of conceptual practice we have thus far identified (e.g., in mathematics: the axioms, the transformation rules, the as-yet-unstated, unproved, unapproached), as well as the social context, the contemporaneous questions asked and the answers given in adjacent disciplines, and so on, is transformed by how things go in the entire process, in the entire mangle. But my way of putting the point makes it clearer that it is the details of some particular mangling that are explanatory, not the mangle itself. The mangle is too abstract and gestures toward too many different kinds of heterogeneous processes to do genuine explanatory work. For this reason it is less suited to do explanatory work than, say, a concept like "social class" — which, despite its well-suitedness, does far less explanatory work than its champions think. (I'll return to this point in a moment.) A concept can be suited for explanation, then, without being especially helpful in explanation.

As a metaphor, the mangle can be extraordinarily helpful in making our explanations *sensitive* to the interplay of all the particularities that make a particular episode in scientific or conceptual practice comprehensible. Adherence to the image of a mangle in which everything participates will make it impossible to allow any explanatory concept to claim "nonmangled" status. Nevertheless, I do not think the mangle is itself explanatory — it points the way toward helpful description, interpretation, and explanation. But not really being an "it," the mangle doesn't explain, nor does it identify a particularly well-defined process. I'm thinking of explanation here in purely pragmatic terms, as involving "finding out" and "making sense," and I'm simply suggesting that it is the details of mangles that will be the important discoveries. Knowing that mangles and mangling are ubiquitous points us in the right directions, but it won't work in any case to say of some process that it is as it is because it participates in the ubiquitous mangle.[2]

Agency. The concept of agency figures crucially in Pickering's account of scientific change, and particularly in his account of pragmatic realism: "Coherences . . . should be seen as [made] things, as actors' achievements, and not as arising naturally and uniquely from the . . . world itself." The relation is one of "made coherence, not natural correspondence."[3] The mind is no mirror, theories aren't copies, and "correspondence realism" is unhelpful in understanding why we speak as we do in scientific and conceptual practice. Practices are guided by human interests and are extended until we come upon resistances and make accommodations, suggest stabilizing strategies, and so on. We bridge, we transcribe, we fill, we confront weirdnesses of various sorts; we adjust, readjust, and accommodate.[4]

The first point I want to emphasize is that the sort of pragmatism — "pragmatic realism" — that Pickering and I (and, I think, Barbara Herrnstein Smith) are attracted to depends crucially on the concepts of *experience* and *agency*, but not a bit on the concept of nonconceptualized *Reality*. But here is the rub. Pickering is fond of talking about disciplinary agency, an important and plausible theoretical posit; he is also drawn, especially in speaking of mathematical practice, to speak of "forced moves." The idea is innocent enough. There are, for example, rules of transformation in place that determine what the properly trained mathematician is allowed to do with a certain well-formed formula. But these "forced moves" are hardly *passive,* as Pickering frequently says. A mathematician could misbehave, refusing to follow the rules; he or she could go to sleep, write a poem, take a walk, or discover nothing on Dublin Bridge and exchange a kiss with his or her lover there instead. This much destabilizes talk of "forced moves" unless it is contextualized to a mathematician actually playing the game and capable of seeing the next possible moves. There is always more than one acceptable move; for example, one could keep rewriting the previous line in a proof if nothing novel came to mind while proving a theorem. So what is "forced" is obscure, and exactly how the individual mathematician is rendered passive even at this stage of transcription is unclear. Indeed, I think it is positively misleading. The mathematician acting within the constraints of disciplinary agency has degrees of freedom, and it is he or she, after all, who *makes* the next move if one is made. There are cases of human activity or inactivity in which talk of "passivity" is genuinely useful, such as when a person is comatose, or frozen, and unable to think of just what action is called for. But I don't think that mathematical practice is one of them. Even a forced move is executed by an agent; it is carried out by an agent with certain commitments to the practices of a discipline and intentions (so far) to abide by those conventions. Normally, the individual who makes a so-called forced move, assuming there are such things, makes the move herself. So talk of passivity seems unnecessarily provocative, taking the decentering of the subject too far, and pressing too hard to avoid appearing in any way aligned with the voluntarist self of the humanist.

The picture with which Pickering and his fellow pragmatists (myself included) will do best is one that locates activity and agency in multiple places. But in so doing, we are not required (nor is it, I think, helpful) to promote a contrasting category of passivity within this heterogeneous space comprised of multifarious types of agents and agency. The best way to sum up my point, perhaps, is to say that I am not denying the usefulness of the category of "forced moves" for certain analytic purposes. What I am questioning is aligning it with *passivity*.[5]

The Metaphysical Impulse. I agree with virtually everything Pickering says in resisting David Bloor's reductive explanation of William Rowan Hamilton's metaphysical meanderings in terms of the too-taken-for-granted concept of "social interest." Social interests and political attitudes are part of the mangle. Whether they are explanatory in this case is, however, another question altogether. Bloor explains Hamilton's gradual shift from a Kantian interpretation of arithmetic, algebra, and geometry to the "formalist" view of Peacock and the Cambridge school, which held that mathematics as a pure science need not be concerned with any particular interpretation of its symbols, in terms of Hamilton's "reactionary" social interests.

Now, something is wrong here. As Pickering points out, Bloor fails to treat the social as part of the mangle, positing it instead as the *cause* of the metaphysical: "The social is the solid, reliable foundation that holds specific metaphysical positions in place in an otherwise open-ended space," as Pickering puts it (70). Pickering rightly suggests that we see the metaphysics of a mathematician "as yet another heterogeneous element of the culture" that scientists operate in and on, along with the social, the material, the political, and so on (74).

Pickering gives an analysis of why Hamilton's metaphysics changed which involves, roughly, the idea that once his geometrizing algebra had taken him into a fourth dimension, he was "flying by the seat of his pants" without much help from grounding intuitions, say, about space, as a good Kantian might conceive of it. This seems plausible — like part of a credible story. But I want to step back to the beginning and recall why Hamilton got into the quaternion business in the first place. Simply put, it was that if — as almost everyone thought until about 1825 — mathematics was the science of quantity, then negative numbers were problematic. But even worse were the imaginary numbers, which involved the square root of negative unity and turned up inevitably when otherwise "reliable general methods were used to solve certain quadratic equations."[6] The problem with the absurd or impossible numbers was assigning an interpretation or a meaning to them that didn't involve an ontological commitment to absurd or impossible things. The meaning or reference of negative numbers and imaginaries (possibly null meaning or null reference) was at stake — a transparently ontological issue.

Kant's great appeal to Coleridge (who sent Hamilton to read Kant), and to most of his subsequent readers, has to do with Kant's interest in grounding mathematics in "sensible intuition" such that the *apparent* necessity and a priori character of mathematical knowledge could be explained. What attracted Hamilton in the "Transcendental Aesthetic" of the *Critique of Pure Reason* was the idea that arithmetic and algebra could be grounded in intuitions about time, and geometry in intuitions about space, that did not depend on empirical knowledge.[7]

Ernest Nagel, in a wonderful essay entitled "Impossible Numbers," asks: "How profound were Hamilton's differences with the linguistic school?" Nagel answers: "He himself indicated that they were not far reaching," and that his "pure Time could be viewed as a *suggesting* science in Peacock's sense, for constructing a symbolic algebra, whose extrasystemic interpretation could then be neglected. It was because the linguistic approach did not afford him an adequate intuition of *number* that he was compelled to dissent from it."[8]

Hamilton wrote in the preface to *Lectures on Quaternions:* "I confess I do not find myself able to form a *distinct conception* of number, *without some* reference to the thought of time, although the reference may be of a somewhat abstract and transcendental kind. I cannot fancy myself *counting* any set of things without first *ordering* them, and treating them as *successive*."[9] And, in an 1846 letter to Peacock, he wrote: "My views respecting the nature, extent, and importance of symbolic science may have approximated gradually to yours. . . . But I still look more and more habitually *beyond* the symbols than they [the Cambridge school] would choose to do."[10]

What I simply want to suggest—and this, I think, is utterly compatible with Pickering's emphasis on metaphysical concerns as part of the mangle of practice rather than oddities to be explained away or some sort of precious afterthoughts (epiphenomena relative to the "really" interesting stuff)—is the *utter cognitive naturalness* of asking questions about the interpretation of mathematical signs. It was such questions that provoked Hamilton and all of his contemporaries to worry about negative numbers and imaginaries; and it was worries about the nature of numbers and magnitudes that provoked Cantor and Gödel and Hilbert and Brouwer to engage in profitable and heated debates about mathematical epistemology and ontology in our own century. Hamilton's metaphysics was hardly an afterthought—it was, as it were, a "beforethought."

What we say about matters of mathematical ontology will be radically underdetermined by how mathematics evolves—as will the norms of mathematical epistemology. But we are creatures of the kind who ask questions that cannot always be answered — questions that regulate our searches, and guide them — and choices must be made about how to guide them. Pragmatic tests are relevant to standing one's ground with a particular form of ontological commitment, or possibly dispensing with matters ontological when it comes to doing mathematics. But it would be surprising if something like the question that bothered Hamilton — "What is a number?"—did not continue to arise, at least from time to time, in response to certain needs experienced by practitioners and internal to various conceptual practices and domains, for mathematics is too impressive an edifice not to wonder *what* it is about — even if at some time we decide that it, or parts of it, is about nothing in particular.

NOTES

1   Andrew Pickering, "Objectivity and the Mangle of Practice," in *Rethinking Objectivity*, ed. Allan Megill (Durham, 1994), 112–13; my emphases.

2   Undoubtedly, someone will ask what I have in mind by "explanation." I hardly have this worked out. But if we are interested in historical or sociological causal explanations, then it seems to me that we want *singular* terms rather than abstract universals in place. If it is some sort of *verstehen* we are after, then my objection to the mangle is not that it is not explanatory in and of itself, but that the mangle will need to be filled out in a detailed way.

3   Andrew Pickering, "Living in the Material World: On Realism and Experimental Realism," in *The Uses of Experiment: Studies in the Natural Sciences*, ed. D. Gooding, T. J. Pinch, and S. Schaffer (Cambridge, 1989), 275–97.

4   Barbara Herrnstein Smith puts a similar point this way: "Beliefs are *modified* in the same ways, through the same general mechanisms, as they are *maintained*. Specifically, our individual tendencies to respond in certain ways to certain perceived cues are strengthened, weakened, or reconfigured by the *differential consequences* of our actions." See her "Belief and Resistance: A Symmetrical Account," *Critical Inquiry* 18 (1991): 125–39; quotation from 135; her emphases.

5   A case where the free versus forced issue came up was in the debate between Gödel and Brouwer over transfinites, nondenumberable infinites, Cantor's continuum, and the like. There was no one disciplinary agent governing both of them. Gödel felt compelled, like Cantor, to allow higher-order infinites so that, for example, the rational numbers had a cardinal of N_0, whereas the real numbers, those on a straight line in euclidean space, had a cardinal of the power set of N_0, that is, $2N_0$. Gödel wrote that Brouwer's intuitionism was "utterly destructive in its results." The whole theory of Ns greater than N_0 was "rejected as meaningless." See Kurt Gödel, "What Is Cantor's Continuum Problem?" *American Mathematical Monthly* 54 (1947); reprinted in *Philosophy of Mathematics*, ed. P. Benacarraf and H. Putnam (Englewood Cliffs, NJ, 1964). Now, there is some sense in which the disciplinary agency within which both Brouwer and Gödel worked was the same: where Brouwer and other constructivists balked was when the forced move, or at least those totally permitted by number

theory, resulted in different order infinites. One way to stop being forced to make Cantor's move, the move Gödel the Platonist heartily applauded, is to restrict mathematics to finitist methods, which will not permit orders of infinity, nondenumerable totalities, and their suite.

6   Ernest Nagel, "Impossible Numbers: A Chapter in the History of Modern Logic," in *Studies in the History of Ideas*, Vol. 3, ed. Columbia University Department of Philosophy (New York, 1935), 433.

7   Although Kant is usually, and rightly, contrasted with Mill when it comes to the foundations of mathematics, there is an important similarity. Mill saw mathematics as enabled by abstractions from our dealings with objects and spaces in the world; Kant saw mathematics as enabled by abstractions from the cognitive conditions which he thought we projected onto experience (out of cognitive necessity). So both, in a sense, based the foundations of mathematics on sensible intuition, one empirical, the other a priori. But they are closer to each other in this respect than either is to a full-blown Platonic realist. See Philip Kitcher, *The Nature of Mathematical Knowledge* (Oxford, 1983).

8   Nagel, "Impossible Numbers," 464; his emphases.

9   Sir William Rowan Hamilton, *Lectures on Quaternions* (Dublin, 1853), 15; his emphases.

10   Robert P. Graves, *Life of Sir William Rowan Hamilton*, 3 vols. (Dublin, 1882–89), 2: 528.

Explanation, Agency, and Metaphysics:

A Reply to Owen Flanagan

Andrew Pickering

I am grateful to Owen Flanagan for his thoughtful comments, and am glad to have the occasion to amplify and clarify some important general points in my essay. Flanagan thematizes three topics: *explanation, agency,* and *metaphysics.* I follow his sequence in my reply.

Flanagan is right to doubt that the mangle is an "it," that is, a particular thing. What "the mangle" refers to in my work is, rather, a complex dynamic process, specifically, the continuously ongoing, open-ended transformation of mutually interactive cultural elements, including conceptual and material practices. In my discussion of the conceptual practices most relevant to William Rowan Hamilton's work on quaternions, I characterize this process as a dialectic of resistance and accommodation.

As this description makes clear, the mangle is not something external to culture that acts on the latter to transform it. Accordingly, I would not say, as Flanagan does, that "social interest and political attitudes are part of the mangle." I would say, rather, that interests and attitudes are part of culture and, as such, part of what gets mangled in practice. It appears to be just this feature, however, that raises the question of *explanation;* for, given certain causal definitions of that concept, if the mangle lacks externality, then it cannot explain cultural transformations. But explanation need not be seen as the indication of linear causal relations. I do claim to provide an explanation of how Hamilton got from complex algebra to quaternions: not, however, by establishing a causal mechanism on the analogy of, say, the collision of one billiard ball with another, but by displaying a graspable structure of dynamic relationships.

This might seem like too attenuated a form of explanation to my fellow snooker players, but it may be the best we can look for. The traditional idea of explanation as a causal account may itself be the problem here. It is certainly a problem in its restrictiveness, which seems to be the legacy of a particular model drawn from scientific practice (and related ideas about how to narrate intellectual history) that we have, I think, good reason to question. I would stress, therefore, that my analysis of Hamilton's conceptual practice is developed precisely as an *alternative* to the sort of "historical or sociological causal explanations" that Flanagan mentions in note 2 of "The Moment of Truth on Dublin Bridge" (with, it seems, a measure of uncertainty on his own part). That was, in fact, the point of the discussion of David Bloor's work in the sociology of science in the last section of my essay: the variables traditionally taken as the enduring, underlying causes of phenomena are, I claim, better understood as being themselves always at stake in practice, always liable to "mangling." Flanagan suggests that the idea of the mangle would be acceptable if only a causal explanation of cultural transformations were added. My own suggestion is that once the idea of the mangle — the ongoing dynamic processes of cultural transformation — is grasped and granted, there is no possibility of, but also no need for, traditional causal explanations.

On the question of *agency,* Flanagan thinks that I have gone over the top. My "talk of passivity," he writes, "seems unnecessarily provocative, taking the decentering of the subject too far, and pressing too hard to avoid appearing in any way aligned with the voluntarist self of the humanist" (85). The problem, however, may be that Flanagan's interpretation of my account of Hamilton is haunted by irrelevant aspects of the traditional humanist idea of agency. According to the latter, what marks human agency or genuine "action" — as opposed to the mere "motions" of, say, machines — is that it is intentional, responsible, and in some sense "free." In these respects, rational human agency is seen as crucially distinct from the mechanical, irresponsible, and involuntary — or, in that sense, "forced" — motions not only of inanimate objects, but also of animals, infants, and adult human beings when, as Flanagan writes, "comatose, or frozen." So, my talk of "forced moves" may have struck him as making Hamilton sound not only passive, but like a mere object or animal, or like someone who is comatose. I think, however, that he missed my point here.

To be sure, a mathematician could always, as Flanagan writes, "misbehave." But that does not contravene — or, in fact, engage — the relevant idea of "forced" in my account. The forcedness of Hamilton's moves was not a matter of something *external to him* moving his hand (or his conceptual machinery) in one direction rather than another. It was, rather, a matter of his own acute, invariant responsiveness to certain features of his world, notably, his disciplinary—discursive, professional, intellectual—world. Thus, if a mathematician such as Hamilton does *not* (will himself to) "misbehave" mathematically, it is not because he is forced by something outside of himself to do something "against his will," but precisely because *he* wants or wills ("voluntarily," one could say—but maybe he can't help it either, given his intellectual history and the apparent alternatives) to remain part of the disciplinary community and discourse of mathematics.[1] Certainly the standard distinctions between what is voluntary and involuntary are unsettled by various twentieth-century understandings of the operations of discourses and disciplines in shaping our conceptions, wishes, desires, and, accordingly, our actions. Posthumanist "decenterings" of the human agent could be seen as just the ongoing elaboration of the implications of such understandings.

Flanagan and I may have a fundamental disagreement about the operations of, in a number of senses, *discipline*. From my own perspective, the "disciplined" human is indeed an algorithmic actor or, as I put it in my essay, a machinelike actor who performs predictably in a given situation. Accordingly, the example that Flanagan gives—the performance of elementary mathematics, such as appears in Hamilton's practice—is not a contrary case, but a paradigmatic example of discipline: strikingly machinelike and algorithmic. It is, in fact, one of the hardest cases imaginable to explain in voluntaristic terms. What one learns in mathematics classes is precisely to manipulate certain elements—say, algebraic symbols—in exactly the same way as everyone else and in no other way.[2] In this sense, therefore, one could say that mathematical practices are "effects" of disciplines—an image significantly different from the humanist/voluntarist idea of *self-originating* human agency.[3] I would add, however, that I do not think that space for voluntary human agency disappears entirely within disciplined practices. Thus I draw attention in my essay to the active, "free" moves that I claim were integral to Hamilton's work, but I also insist that these were constitutively intertwined with disciplined "forced" moves in the sense outlined in the essay and amplified above. The identification and analysis of this intertwining in Hamilton's conceptual practice is what makes the explanation in my essay posthumanist, which is to say, neither humanist (centered on the voluntary human agent) nor antihumanist (centered on discipline). I must reject Flanagan's suggestion, then, that I took the decentering of the human subject too far. I took it, I would say, just far enough to enable me to make sense of conceptual practice within the larger explanatory framework of my account of the mangle.[4]

The last topic that Flanagan discusses is *metaphysics*. His observations here are interesting, and I have little quarrel with them substantively. I would, however, question some of his locutions, such as his suggestion that "we are creatures of the kind" who necessarily or naturally engage in metaphysical speculation (88). I can remember learning about complex numbers, but I cannot remember ever wondering about their metaphysical foundations — not, that is, until I tried to understand Hamilton, and then I did wonder a lot about them.[5] I am also not sure what Flanagan means by the "utter cognitive naturalness" of Hamilton's interest in "the interpretation of mathematical signs," which sounds as if ultimate meaning or reality were something we are all born to be concerned about (87). Flanagan seems to be suggesting here that there are innate interests or universal metaphysical impulses that "regulate" and "guide" our "searches" independent of particular cultural or disciplinary practices. My own inclination is to see metaphysical worrying about foundations as an *option* for conceptual practice, but as neither its enduring, underlying cause nor the product of impulses that are any more utterly natural than they are utterly cultural.

NOTES

1   Of course, Hamilton seems to have been responsive to some especially subtle configurations of his conceptual and intellectual world as well as more capable than most people, including most other mathematicians, of entertaining alternative configurations and pursuing their conceptual implications and technical viability. Such unusually subtle responsiveness and energetic innovativeness is often referred to as "genius," "creativity," or "imagination," but its operations can, I think, be analyzed, which is what I attempt to do in my essay.

2   There are already machines that can perform what I describe as passive or "forced moves" in mathematical practice: computers running *Mathematica*, for example. Such machines do not, however, have the capacity to simulate what I call "free moves."

3   It is presumably because I too have been disciplined that I can make sense of Hamilton's writings and that other, similarly disciplined persons can check the technical details of my analysis. It is hard to imagine how mathematics could exist without this machinelike quality of mathematical performances. Flanagan does not offer an account of the operation of disciplines in accordance with a voluntaristic conception of human agency (at least, not in his commentary here), but it would have been useful to have had one.

4   For an antihumanist account of mathematics, one can consult the great mathematician Alan Turing; see, for example, Andrew Hodges's wonderful biography *Alan*

Turing: The Enigma (New York, 1983). The "Turing machine" was initially a conceptualization of the work of human "computers" reduced to its bare bones; only later did a real machine, the electronic computer, come to be regarded as an instantiation of this concept. The importance of "slaves" in Turing's thought on mathematics is also worth pondering. (The index to Hodges's book lacks any entry for "slaves," but see, for example, 211, 278, and 292.) It is clear, though, that Turing's vision cannot accommodate the processes of cultural transformation addressed in my essay.

5　Hamilton's mathematical contemporaries in Cambridge were not especially interested in metaphysical foundations, and, indeed, many of them were rather intolerant in that regard: they recognized a metaphysical impulse when they saw it, but, far from acknowledging it as the reflection of universal human nature, condemned it as a nasty habit. Hamilton's differently disciplined *philosophical* contemporaries, of course, were likely to have viewed such interests and impulses in other ways.

Agency and the Hybrid *Collectif*

Michel Callon and John Law

COULD non-humans ever be agents? That is the topic of this paper. It's a hot topic. One that exercises philosophers, sociobiologists, and theologians. Not to mention sociologists, sciencefiction writers, and those who work in science, technology, and society (STS). It's a hot topic because it sometimes seems that there are all sorts of non-human entities, such as cyborgs, intelligent machines, genes, and demons loose in the world. Along with ozone holes, market forces, discourses, the subconscious, and the unnameable Other. And, or so many claim, such non-human actors seem to be multiplying. For if angels and demons are on the decline in the relatively secularized West, then perhaps robocops and hidden psychological agendas — not to mention unnameable Others — are on the increase.

We're interested in non-human agency because we work in STS. This has a particular problem: one of its primary concerns is the study of machines and devices. Some of these seem to be very clever, and many of them seem to operate as if they were agents. Therefore, it is no surprise that in STS there are substantive debates about whether computers, robots, or, for that matter, animals, are "really" agents or not. Which is the reason why we were interested when we saw a flier for a conference on "Non-Human Agency: A Contradiction in Terms?"[1]

JOHN: "Non-Human Agency: A Contradiction in Terms?" The title sounds all wrong to me.

MICHEL: Yes. That's right. And we're not alone in thinking so. For instance, Steve Brown and Nick Lee make the same argument.[2] For to ask us whether "non-human agency" is "a contradiction in terms" is a trap. One: it asks us to assume that there are indeed two classes, human and non-human. Which we don't. Two: it asks us to assume that humans are special and particular. That they are indeed agents. So any notion of agency that we might end up with will derive from some idea about the character of human agency. And three: it asks us to consider, like good liberals confronting the lower orders, whether the franchise should be extended. Granted to non-humans.

JOHN: Okay. I agree. But Steve Brown and Nick Lee are also complaining about actor-network theory (ANT).[3] For instance, they intend the last point as a criticism of ANT.

MICHEL: A criticism?

JOHN: Yes. Because ANT is a body of writing which says there are lots of entities out in the world. But most of them don't get properly included in the stories told by social scientists.

MICHEL: How so?

JOHN: Well, either they are ignored altogether — technologies, animals, environmental processes, angels or fairies, they are simply absent from social science stories — or they get, as it were, passive, walk-on parts. They are treated as if they were simply the instruments, perhaps the collective instruments, of people and their actions.

MICHEL: Well, I don't think it's as simple as that. Technologies are sometimes said to drive the world.

JOHN: Okay, but even so, they're said to be different, different in kind from people. They're not actors in the proper sense of the word, the sense that you'd use about conscious human bodies. And Steve Brown and Nick Lee are saying that ANT is about extending the intellectual franchise. To include animals and cyborgs.

MICHEL: And they're right. As Vicky Singleton and Leigh Star have noted, ANT tries to assimilate the Other.[4] To efface it. To conceal ambivalences. Which means that we not only have to escape the terms of the debate: this distinction between human and non-human. We also have to find a way — no, better, let's say, many ways — of transcending the egalitarian panopticism of liberalism.[5]

So we start with this assumption: that non-human agency is not a contradiction in terms. That it is possible — indeed interesting — to say that there are non-human agents. That, indeed, there *are* non-human agents. Though even putting it this way is a bit problematic, since it assumes that there are identifiable "humans" and "non-humans." But there's a problem of argument too. Namely, that debates about the status of agency are metaphysical. Are "humans" "like" "non-humans," or not? This is undecidable. Or perhaps it can sometimes be decided, but only locally. So we can't in general *prove* that "humans" are like "non-humans." Or that some "non-humans" are agents. If we tried, we'd be wasting our time. All we can do is make stories which suggest that if you don't make such assumptions, then revealing things may happen, theoretically and empirically. And that is what we'll do. We'll adopt a methodological point of view: that it is interesting to leave the question of agency open.

ACTION DERIVES FROM THE "COLLECTIF"

Andrew is the managing director of a very large laboratory.[6] He's able, he's personable, and he's fierce. He's an actor. An agent. But let's look at some of the things that he does. For instance, he talks; he talks with subordinates. And he also gathers intelligence from outside. An example: he visits "head office," where they make decisions about the future of the laboratory. And, at the same time, he lobbies. So half the time he's not sitting at his desk. Indeed, he's not in the laboratory at all. He's at head office. He's negotiating. He's collaring people in the pub to bend their ears and catch the latest Whitehall gossip. Or he's in London, attending meetings in smoke-filled rooms. Comparing brands of malt whisky with the powerful in the land. He's out there "waving the flag," as one of his subordinates puts it.

Now let's try a thought experiment. Just imagine what would happen if they took away Andrew's telephone and his fax machine. If they blocked the flow of papers and reports. Imagine what would happen if they shut down the railway line to London and stopped him from using his car, so he couldn't get to the smoke-filled rooms. Then imagine, also, that his secretary were to disappear. And his room, with its conference table, its PC and electronic mail, were to vanish. Imagine what would happen if other people in the laboratory started to ignore him. Or to treat him as if he were the messenger bringing 'round their letters and paychecks.

Call this the baboon experiment.[7] An experiment where, as in a Paul Auster novel, the materials of sociality are progressively withdrawn until the person is sleeping rough in Central Park.[8] Under such a regimen of deprivation, would Andrew still be an agent?

Undecidable question. That's the problem with "just imagine" stories. But it's also the problem with metaphysics. Is Andrew still Andrew if we reduce him to the status of a tramp? Or, for that matter, to a heap of rags and bones? It's possible that there is something "essential" about "Andrew the person" that means he's still an actor. For instance, until he dies. Or for all eternity, in the form of his immortal soul. That's the dualist metaphysic. Two classes of things: human, and non-human. But there's an alternative, an alternative which suggests that Andrew is an actor because he is a particular kind of emergent effect, an arrangement of bits and pieces. No doubt, at least some of the time, this has something to do with his body, but it also has to do with a whole lot of things that lie beyond his skin. That's the point of the baboon experiment. It illustrates the *possibility* that agency is an emergent property. That to be an agent like a managing director is a form of action which derives from an arrangement. That, *by themselves*, things don't act. Indeed, that there are no things "by themselves." That, instead, there are relations, relations which (sometimes) make things.

JOHN: Okay, Michel. Here's the big question. Are you an agent?

MICHEL: Oui. Effectivement. Pour l'instant j'agis comme quelqu'un qui est doté de ce que vous, les Anglophones, vous appelez "agency."

JOHN: So why is that? How come you're acting as if you've been endowed with agency right now? In what sense?

MICHEL: C'est parce-que pour l'instant—j'insiste sur ce point, *pour l'instant*—je re-présente un collectif hybride. Et, simultanément, j'appartiens à un collectif hybride.

JOHN: Let me translate. You're saying that you're an actor here and now because your voice re-presents a "hybrid collectivity." And because it is a part of it.

MICHEL: Non. Tu as tort. Je ne re-présente pas une *collectivité*. Je re-présente un *collectif*. Les deux notions sont très différentes.

JOHN: I'm losing confidence. I thought I could translate for you, but you're saying that I'm getting it wrong. You're saying that what you call a *collectif* is not a collectivity.

MICHEL: Oui, c'est cela.

JOHN: Let me guess. A *collectif* is an emergent effect created by the interaction of the heterogeneous parts that make it up.

MICHEL: Oui, c'est à peu près cela. Ce n'est pas du tout un ensemble de personnes déjà-là et qui décident de se lier par une organisation commune.

JOHN: So it's the *relations*—and their heterogeneity—that are important. Relations which *perform*. Perform agency, at least in this case. Perform a collectivity to which your voice belongs. And for which, perhaps, some of the time, it speaks.

MICHEL: Nous y voilà, mon cher collègue! Nous formons un collectif au nom duquel nous agissons. Tu me re-présentes autant que tu le re-présentes.

JOHN: Okay. I can speak for you, even belong to you, so long as I remember that a *collectif* is not a collectivity.

Perhaps, then, the lesson comes from semiotics. At any rate, it's a useful tool.[9] If we want to imagine that there are relations. *Collectifs.* Not things in themselves. Or even special and distinct classes of things that have relations. Like humans and their linguistic relations. Just relations. Links. Interpenetrations. Processes. Of any kind.

ACTION MAY BE SEEN AS MATERIALLY HYBRID

JOHN: Earlier, you said that you were an agent because your voice re-presented, at least for a moment and in those particular circumstances, a hybrid *collectif.* And because it belonged to this *collectif.* Now I've got a couple of thoughts.

MICHEL: Yes?

JOHN: Well, first, I'm not sure that I think all hybrid *collectifs* are agents. I guess that there's something else going on too, something that turns hybrid *collectifs* into agents. Sometimes.

MICHEL: Yes. Okay.

JOHN: But we can come back to that later, this question of distribu-tion. Because what I'd really like to do is to ask you a personal question. About the nature of the "Michel Callon *collectif.*" About its organization and its components.

MICHEL: That's not a personal question.

JOHN: Why not?

MICHEL: It isn't personal because the syntax of the "personal" im-plies precisely the kind of dualism that we need to try to avoid. As if there were personal, private, specifically human things. And then other kinds of things. That's the whole point of speaking of the *collectif.* That it erodes divisions and distinctions that are said to reside in the order of things.

JOHN: Complicated! Complicated because I guess that, in our *collec-tif* monisms, we tend to perform dualisms. That's what Norbert Elias suggests, anyway.[10] But let's agree, and try to pose this non-personal question again. So what's the character of the *collectif* that your voice re-presents?

MICHEL: Oh. I don't know. Lots of texts. Read and written. A lot of notes. A lot of conversations. A whole lot of tapping into computers. My Macintosh Powerbook and the electronic mail. Some lines of argument.

Re-presentations that emerged from your mouth.

JOHN: You're claiming, now, that *you* include *me?*

MICHEL: No. But these words include some that came from your mouth. And your fingers as you sat at the computer and raced against the rapidly approaching deadline.

JOHN: Tiens. Je . . . pardon . . . nous ne pouvons plus vous tutoyer. Vous êtes pluriel.

MICHEL: In English, you sometimes call it the decentering of the subject.

> One day Andrew called a meeting of his top managers. It was a crisis meeting. The laboratory was investing in a big instrumental project. Its success was vital to the future of the organization. It was the opportunity of the decade, the "flagship project." But Andrew scented trouble. Why?

> The answer was that there didn't seem to be enough people working on it. The laboratory had set a target: so much effort had been planned and forecast. But in practice the laboratory wasn't hitting that target. And since there wasn't room for any slack, if the amount of work actually done fell short, then the project would be delayed. Which meant in turn that the laboratory would look inefficient. So this was Andrew's problem. He wanted to review what was happening. And he wanted to find ways of increasing the amount of effort going into the project if it turned out that this was needed.

Big questions: How does Andrew *know?* No, better: How did Andrew come to know, at this particular time and in these particular circumstances, that there didn't seem to be enough people working on it? How was Andrew able to think this? Then and there? What was the process that generated this "thought?" Let's go on with the story.

> In front of him Andrew has some figures. These figures come from the "manpower booking system." That is, they come from a set of relations and processes, spread out through space and time. Which end up, for a moment, on a desk. And "re-port," bring back, the "man-years" devoted to each project. But the man-years put into the flagship project are way, way down. That is, they are way, way down compared with the plans and projections. Which makes Andrew suspicious. Perhaps things are going well. Perhaps, just possibly, they reflect unanticipated productivity. But there's the other possibility, let's say, probability: that trouble is brewing. That the work is not getting done.

So how does the manpower booking system work? How does it report, bring back? A partial answer: every month everyone in the laboratory fills in a time sheet. They fill in a series of codes which correspond to particular projects, half-day by half-day. And then they hand the forms in. And down in finance these are collated and corrected (whatever that might mean). And guesstimates are made for those that haven't been filled in properly or returned at all. And the results are typed into a computer, which does the arithmetic. And the result is printed out in the form of a spreadsheet.

Pretty mundane stuff really. Every organization does it. But let's make a brief list of the kinds of *materials* involved in this process. One: there are people. Apart from Andrew himself, there are many others. For instance, the people who fill in the time sheets. And those who collect, collate, and correct them. Two: there are papers. For instance, there are time sheets and computer printouts. And so on. And so on. And three: there are devices and other material products. For instance, computers and sheets of blank paper. Not to mention telephones and printers. Or notebooks and pencils.

How long do we want to make the list? Do we want to include what we sometimes call "the social" as a distinct category? If so, then we need to add organizational arrangements.[11] And how about the symbolic? Do we want to add symbols and arithmetic?[12] And how about architectures?[13] The answer to the question of how long the list should be is a matter of taste. For this is the argument: it is plausible to say that Andrew is able to act as a strategist because there is a *collectif*, a *collectif* of materially heterogeneous bits and pieces. Because there is a *hybrid collectif*.

That's our suggestion. The dualist view is that there is, as it were, a "ghost in the agent." That she has a soul, a moral sense. Or that she's a member of a language community. That she is ultimately different in kind. Fine: that's a form of metaphysics. We can't disprove it. But what we'd like to try to do is to suggest the plausibility of work based on an alternative metaphysics.

MICHEL: You know, we've given too much away.

JOHN: Given too much away? How so?

MICHEL: We've said, "It is plausible to say that Andrew is able to act as a strategist because there is a *collectif*, a *collectif* of materially heterogeneous bits and pieces."

JOHN: Yes?

MICHEL: Well, I'm not so keen on the "because."

JOHN: Because?

MICHEL: Because it sounds, I don't know, far too definite. As if Andrew were caused to act by all this information. Or perhaps better, as if he were separable from the role he plays as manager.

JOHN: And?

MICHEL: Well, he is, to be sure. Sometimes. But to put it in that way is to give too much away to dualism. Why didn't we say, for instance, that the *spreadsheet* is able to act as a strategist because it is a *collectif*?

JOHN: Because no one thinks that spreadsheets act like strategists.

MICHEL: That's the point. This language still tends to perform dualisms. Or at any rate, distinctions given in the ontological order of things. It still tends to perform people as different in kind. As those entities that have the right to speak and vote. And to make strategies.

JOHN: So the problem is partly this. It's to make sense of the way in which divisions — if you like, dualisms, such as the gap between human and non-human — are generated. Out of partial similarities.

MICHEL: That would be the difference between a theory of the *actant* network and the theory of the *actor*-network.[14] All actants are created equal. But actors have distributions thrust upon them.

At this point we want to make a few comments about attribution. Let's try the following as a working hypothesis.

In many parts of contemporary Western culture, something is treated as an agent or, at any rate, as a candidate for agency if it performs, or might perform, two great classes of conditions: *intentionality*, and *language use*. A note on each.

Intentionality: Let's say that this is a family of notions. We don't legislate, but simply observe, that these notions seem to include the idea that agents are those entities able to choose, to attribute significance to their choices, to rank or otherwise attribute "preference" to those choices; the idea that agents have goals, that they are able to monitor and assess what is going on, that they are reflexively able to make connections between what they take to be going on and those goals; and, the idea that, at least some of the time, agents are able to intervene — to act — in order to (re)create links between their goals and the actions that they cover.

SYNTAXES OF ATTRIBUTION

JOHN: This is beginning to make me most uneasy.

MICHEL: Why so?

JOHN: Two reasons. The first has to do with the notion of intentionality itself. My colleague Rolland Munro has "reminded" me that intentionality comes before "intentions." It's a ground for action in the phenomenological tradition: the ground of "aboutness"; so it doesn't have to do with being about this or that thing in particular. Or about intentions.

MICHEL: Okay. So since we're not trying to build on a phenomenological ontology, let's say that we're talking about "intentions," or "intention," rather than intentionality.

JOHN: Well, at least that's clear. I don't think we're wanting to press a phenomenological ontology. But the problem is that it leads us back to the tricky dualism between subject and object. Which is the second problem, one posed by Annemarie Mol. She asks: What's the status of our description here? Are we "following the actors"? And offering a fancy version of *verstehende* sociology? A sociology that deals with actors' meanings? Like that of Weber? Or are we offering our own description? I have to say that at this point I'm not clear about that. I thought we wanted to break out of a sociology that resided in a discursive "order of things." But to speak of "attribution" and then to link it with "intentions" sounds like a move in *verstehende* sociology to me. It sounds as if it has to do with meanings.

MICHEL: Okay, but what's the importance of deciding this for the question of agency?

JOHN: It's simply this: that if we follow a *verstehende* sociology we end up, in one way or another, with a definition of agency which, while varying between contexts and cultures, is always a definition that reflects what people-agents are prepared to count as agents. So it's speciesist; it falls back into the trap posed by the question with which we started: whether (or not) "non-human agency" is "a contradiction in terms."

MICHEL: Okay. I see the point. But I think I see a way around it.

JOHN: What's that?

MICHEL: I think we're going to have to move from talk about representation (which is, after all, quite consistent with *verstehende* sociology, with how people "represent" things to one another) to talking of *signification*.

JOHN: You speak in riddles.

MICHEL: Have faith, my dear colleague! We'll get there in the end.

To say it simply, intention (not intentionality) has to do with *reflexive teleology*.[15] Entities which perform reflexive teleology are candidate agents, especially if they also use *language*. Which brings us to the second great Western condition for candidate attribution to the class of agents. Again, we don't stipulate. We merely note that agents are often said to be symbol-users; they are said to be able to make connections between the statement of rules and their "proper" performance, which involves, in turn, the notion that terms are multi-interpretable, and that proper actions are not given in their verbal description.[16]

Let's repeat. We offer these remarks as suggestions. Revisable suggestions. We don't want to legislate what should count as an agent and what should not. And we're also aware that in some measure we're eliding the problem, by slipping between a *verstehende* understanding of agency and one that is quite different — perhaps semiotic. But matters are complicated in other ways too. They're complicated because the grounds of attribution, or better, the acts of attributing, tend to have effects. They tend to *perform themselves* in the shaping of the *collectif.* They tend to produce and reproduce agents. Agents that perform themselves as reflexive and teleological language-users. Which means that as things stand at present, the hunt for agency is often going to take us, empirically, in the direction of intentions and language-use.

MICHEL: But there are other things going on. Counterveiling forces.

JOHN: How do you mean? Like what?

MICHEL: Teleologies that don't map onto standard material packages like humans implicated in a language.

JOHN: Like computers, you mean?

MICHEL: Yeah, or organizations.

JOHN: Perhaps, then, this is the place where we start to move from *verstehende.* Where culture starts to corrode its own order of things.

MICHEL: What are you thinking of?

JOHN: What you've just said. If we're willing to detach the idea of reflexive intention from a necessary link to understandings shared or assumed by human actors, then there's nothing to stop us "misapplying" them, for instance to machines.

MICHEL: Okay. I see what you mean. And perhaps this is made all the easier as a result of science and technology. Not so much that these generate entities that are heterogeneous — for there were always entities that were heterogeneous.[17] But that it does it so fast — that so much is changing so much of the time — that distinctions and attributions which more or less stabilized themselves at the time of the Enlightenment are starting to lose their ontological status. Anomalies keep on popping up. Questions keep on posing themselves.

JOHN: A step change in the pace of innovation. A new global hyper-reality. Possibly. Though I'm not very persuaded by the theoretical significance of these historicist arguments.[18] But never mind.

Andrew's figures are very particular. They aren't just any old figures. This is because they allow him to see what is going on. No,
better, they are just one of the many processes that allow it to be
said that Andrew can "see" what is going on. For there are questions, for instance, to do with the attribution of expertise. The
managing director has the qualifications to "see," whereas the clerk
who prints the figures out does not.

So let's phrase that more carefully. The figures participate in the
creation of a place of visibility. They participate in the act of visibility. Things, processes, entities, relations that are spread across
time and space are brought together. We're witnessing a putative center of translation. A panopticon.[19] So this is the materialist
point: the creation of Andrew as manager can't be detached from
the business of creating the right kind of paperwork. And, to be
sure, other materials too.

That's what we mean when we speak of a hybrid *collectif*.

For what we call "a manager" (which is, after all, a particular kind
of agent) is created in part in the process of generating paperwork, paperwork which allows it (the manager, the *collectif*) to see
if things are going wrong. And which offers it the possibility of
acting to try to put matters right if problems seem to be arising.
Such an entity, then, is brought into being to the extent that the
paperwork tends to generate a place of discretion. Which is why,
to be sure, the manpower booking figures are so very important.

Something else: we've got a *material difference*. That is, we've got a combination of two different kinds of materials: a human body, on the one
hand; and a text, on the other. Perhaps we should say that the combination of the two generates a material gradient. For since the text
re-presents other, less malleable and tractable events — events which
cannot be seen in a single place — the text creates a space where options
may be exercised. Informed options. A *discretionary space*. We're dealing, then, with a mode of ordering which defines and distributes the
character of persons and papers. *Which projects itself onto and through
materials of all kinds.* And necessarily does so in a heterogeneous manner. For persons are projections of a logic which generates them (or, at
any rate, some of them) as discretionary beings endowed with powers
of foresight and capacity for choice. With intentions. But this isn't pos-

sible without the relevant paperwork. So papers, in turn, are projected as little vignettes, summaries, simplified concentrates which constitute the heart of things and thus act as an aid to discretion.[20]

And similar arguments may be made about other materials. For the papers don't emerge from nowhere. Computers and computer programs project and carry forward the strategic logic by adding, subtracting, dividing, and averaging. And so on — for all are effects and projections of the same kind of strategic logic. A logic of agency. A logic of intentions.

So this is a distribution. A distribution of roles between materials which also produces those materials.[21] Projects itself onto, into, and through those materials. And, at least to the extent that it loops back upon itself, goes on reproducing that distribution and those materials. Goes on reproducing that particular form of agency.

MICHEL: I've been meaning to ask you. If the word "collectif" doesn't translate into English, then the notion of "agency" doesn't translate into French. So what does it signify?

JOHN: I was afraid you'd ask me that! I think that's what we're discussing. And I'm afraid it's contested terrain. It drags us back to the discussion we've already had about *verstehende* sociology. But here's a start. If you look at the beginning of Monica Casper's paper for the Surrey conference, you'll find that she includes a dictionary definition.[22] It's a definition that contrasts the "volitional, purposive" character of human activity — that's the agency bit — with the constraints and determinations imposed on activity by social structure.

MICHEL: But that's incredible! Like the framing of the conference question, it rests on a dualism. Agency versus structure. Another either/or. Another way of dividing the world into two classes: subjects and objects, society and nature, people and things. So only *people* have agency. That's it. End of discussion.

JOHN: Hang on a minute! Like Lévi-Strauss, you only see bifurcations and oppositions. But it's really more complicated than that. Even in the English language.

MICHEL: Ah, thank God for the infinite subtlety of the Anglo-Saxon mind!

JOHN: Hmm. . . . "English-speaking" doesn't mean "Anglo-Saxon." A French misunderstanding with racist overtones. But never mind. Because sometimes it is said that structure *enables* agents as well as *constraining* them.[23] But I think we can do better than that. I think — and this is the place where we start to prize ourselves loose from the clutches of *verstehende* sociology — we can say that agency comes in a variety of forms. Let's use the language. Hybrid *collectifs* come in various shapes and sizes. Some are strategic, and some aren't. Some are intermediaries that just pass on messages, and some aren't.[24] Some seek to boldly split infinitives where no man — excuse me, no entity — has done so before. Some are unpredictable and capricious, like the Greek Furies or Shakespeare's Ariel. And some are decentered. All over the shop.

MICHEL: All over the shop? What do you mean?

JOHN: Fragmented. Dispersed. In a state of disorder.

MICHEL: Fragmented. I don't like that!

JOHN: Why not? Are you still attached to the managerialism of ANT?

MICHEL: Mea culpa. I have sinned! No. On the contrary. I think that the language of "fragmentation" is indeed, as you put it, managerialist. It implies that agency is sometimes a broken whole. That the heterogeneous engineering hasn't quite worked. That it's failed. That if things were better everything might fit together.

JOHN: So agents are, perhaps, sometimes decentered?

MICHEL: Sure: that's what's so interesting. Once we divorce the notion of agency from the panopticism of strategic intention — once we no longer imagine that the former necessarily implies the latter — then we open up a whole field of empirical and theoretical possibilities.

JOHN: Good. At last!

MICHEL: At last?

JOHN: Yes. At last, we can participate in the huge literature on conflicting identities. On decenteredness. On the connections that are multiple and partial.[25] To appreciate that we are not alone in STS, but that we're partially connected!

So we're saying two things. First: that *the hybrid collectif sometimes generates discretionary places*. And second: that agency is (usually) attributed to a particular part of the hybrid *collectif*. For instance, to Andrew and his management team. So it's like this. Andrew and his colleagues are *created* as strategists, that is, as a specific class of agents in the hybrid *collectif*. They are generated by it. And *then* it is said of them that they *are* the agents. But it's complex. Because if they were replaced by Tom, Dick, and Harry, if agency were attributed to these instead, then there'd be another, somewhat different hybrid *collectif*, a somewhat different strategy. It would be different because they'd bring other *collectifs*, somewhat different arrangements of materials, with them.

JOHN: Okay. But there's a problem. We're talking of attribution. But *who* or *what* does the attributing? We still haven't answered this question! We're still not clear whether we're dealing with a class of embodied language-users — people — who graciously award or withhold the benefits of agency to non-humans. Or whether we're offering a different kind of diagnosis: one that allows us to attribute agency in ways that are, as it were, orthogonal to human understandings about agency.

MICHEL: Okay. I agree. But let's deal with this question in two steps. First step: it's obvious that there's a strong tendency to attribute agency to persons. To human bodies or human minds. But there's something else going on too. Second step: for, less obviously, there's also a strong tendency — let's put it no more strongly than that — to attribute agency to *point locations* — places or points that last, that keep on going for a time (whatever that might mean).

JOHN: So fields, diversities, processes, or areas — whatever the metaphor may be — agency isn't attributed to these.

MICHEL: Exactly so.

Listen to this.

> We're in a committee. It's the safety committee of Andrew's laboratory. And there's an argument going on about the safety of a particular installation. The equipment makes use of very powerful X rays, and everyone is agreed about one thing at least: those X rays shouldn't be mixed with people. But how to keep them apart?

> There are two schools of thought. The first says that things are reasonably safe as they stand. And the second disagrees. But, says the first, we've operated our equipment like this for the best part of ten years and no one has experienced any harmful radiation. We've never had an accident. Well, says the second, that's true, but it's a fluke. No, it's not, says the first. It's not a fluke because we've got these detailed written rules. Look, they define exactly what people may or may not do. For instance, they say that no one may enter the dangerous area until they've gone and signed out a key from the safety officer, and the safety officer cannot sign the key out to anyone until the X rays have been turned off.

> Hah! says the second. That's precisely why we say you've been so lucky. The rules say what should be done. We agree. We can see them. They are written down. But it's only a matter of time be-

fore someone forgets to follow them properly. For, apart from the rules themselves, there's nothing to stop someone doing the wrong thing. Getting hold of the key in the wrong way. Which means that this equipment is an accident just waiting to happen. So we've got to build a proper interlock, something that turns off the X rays automatically if someone unlocks the door. Only then will we be really safe.

Our argument is that this story strains to distribute agency to places, to points. Despite the fact that we're dealing with multiple materials, it isn't *interactions*, networks, or fields that are being endowed with agency, intention, or the capacity to use language. It's *singularities*. Like people, or, just possibly (though this isn't clear), machines. And this seems to be usual in the discourses that we know: agency is said to be—is performed as—localizable.[26]

MICHEL: What a very strange finding. I wonder why that is?

JOHN: I'm tempted to read you a lecture about the Enlightenment project.

MICHEL: You should resist the temptation! I'm sure it goes back much further than the eighteenth century. At least in some parts of the West.

JOHN: Yeah, I agree. But it's interesting, isn't it, to think about the kind of argument made by Richard Sennett?[27] The kind of argument which says (of people, to be sure) that there has been a secular shift away from concern with contexted performance—a "good act in the circumstances"—toward obsession with essence—the "core" of the person. So that, for instance, we are more likely to judge our politicians by asking whether they are "good" than by looking at what they do. Hence the obsession, for instance, with sexual impropriety.

MICHEL: An obsession peculiar to Anglo-Americans!

JOHN: I'm not so sure any more. The infection seems to be spreading. But anyway, it makes me wonder whether it's not more than coincidence that it's French philosophy, with its poststructuralism, which has most ruthlessly tried to decenter the subject. Though the argument has been picked up and pressed home by radical English-language thinkers.

MICHEL: Who are you thinking of?

JOHN: Those—especially radical "postmodern" feminists—who've tried to find a way around the fissiparous politics of essentialism. Those who've tried to argue that the way forward is to make alliances, partial connections.[28]

So, in the process of attribution, there is a bias toward singularity. A bias from which small parts of social and political theory — and practice — are attempting to extract themselves. But there is something else going on too. And this, we suggest, is the moment where we finally distinguish ourselves from what is most important about *verstehende* sociology. And this has to do with the question of language.

MICHEL: I guess we agree that linguistic sense-making is an important part of the process of attribution. Certainly, that's the way we've been speaking of it. We've mentioned this several times, but now we need to rehearse the argument one last time. The linguistic view is that only humans (and, possibly, certain marginal classes) are endowed with language. That is, only humans are able to describe the world, to share those descriptions, and to argue about them. Other entities are deprived. They haven't got the ability to do this. For example, the safety officer insists that interlocks follow the rules. Willy-nilly. Whereas people may, or alternatively may not, choose to do so. Or do so in the way intended by the safety officer. People are Wittgensteinian. Interlocks are not.

JOHN: Okay.

MICHEL: I think this argument is wrong. I think it's wrong because it rests on and reproduces a distinction between language and the world. The world is treated as a referent. It's *outside* language. It's outside *all* languages. And languages "represent" that reality. They represent it in a more or less transparent manner.

JOHN: I'm worried about saying, as you do, that an argument is wrong. If knowledges are local, then arguments are right or wrong. But only locally. But never mind, because I think I hear you affirming a preference for continuity and flux rather than distinctions given in the order of things.

MICHEL: Okay. That's a way of putting it. But I feel more combative about it than you. I want to add that this understanding of language has been under pressure for at least twenty years. The idea that language is a medium, a more or less transparent medium, for a reality given in the order of things no longer works. At least in the places I inhabit. Instead, it's come to be seen as opaque. It's come to be seen as a process which builds readers and worlds all together, at the same time.[29]

JOHN: That's the semiotic turn.

MICHEL: That's right. But you can see it at work, empirically. For instance, you can see it at work in the safety committee. This is the argument: there isn't a context *within* which, and *about* which, the talk about safety takes place. There is no useful distinction to be made between content and context. The two go together. They're built up together. And the distinction, if distinction there be, is fabricated, all in one go. Discursively. The machines, the X rays, the people, the speakers — all of these are built up together, as a package. And the distributions and attributions — these, too, are all part of the package. Which is how we get to the slogan, the rallying cry: "Everything is text; there is only discourse."

JOHN: I repeat: this is the semiotic turn. Distinctively European. That is, *Continental* European. Most English speakers — except, perhaps, for those influenced by Michel Foucault — have never felt really comfortable with this argument.

MICHEL: That's right. But there's an analogous problem for the Continental Europeans. And the problem is this: the assumption that, because texts are opaque, there is *nothing* beyond the text. No outside. Nothing else.

JOHN: So you want to push us back to a distinction between text and context?

MICHEL: No. Not at all. I'd like us to imagine chains or networks of re-presentation. This is the point. We need to drive a wedge between re-presentation and language. Between making present again, and its linguistic form. Sometimes re-presentation comes in the shape and form of words. But often it does not. That's the virtue of all this work in science studies about laboratory science. Chains of re-presentations *may* lead to words. May come to take the form of words. But they may equally well lead to, or take the form of, technical objects — for instance, instruments, or diagrams, or skills embodied in human beings.

JOHN: And architectural arrangements. Have you heard about the Frank Gehry house in Santa Monica? A three-dimensional re-presentation. One that can't, or so Fredric Jameson argues, be reduced to the two dimensions of a photograph.[30]

MICHEL: A nice example. So the argument runs like this. There isn't a reality on the one hand, and a re-presentation of that reality on the other. Rather, there are chains of translation. Chains of translation of varying lengths. And varying kinds. Chains which link things to texts, texts to things, and things to people. And so on.

JOHN: Chains which *make* the things, the texts, and the people. Chains in which the making and the re-presenting cannot be distinguished. Except locally, and for certain purposes.

So this is the argument: We'd like to avoid the trap of imagining that discourse is a closed universe, for that way leads to idealism, and to the famously arbitrary nature of the sign. Instead, we'd like to broaden this notion of signification, or re-presentation. We'd like to broaden it to allow for the possibility that re-presentations come in all kinds of more or less hybrid material forms. Some of which have little to do with language. Some of which, indeed, are unspeakable. *Cannot* be said.

The point is illustrated in an entertaining paper by Malcolm Ashmore.[31] He tells about the interaction among an adult feline ("Smith"), a catflap ("Catflap"), and an adult human ("Alison"). What he does is to try to press a symmetrical account of a series of interactions among these three entities. Or, more precisely, he tries to write a *series of accounts*, with motives, intentions, plans, preferences, and interests, as if they were written from the point of view of each of these entities. But, or so it turns out, it's quite difficult to write a convincing narrative of some of the events from, say, the standpoint of Catflap. But why this might be so is now clear. It's because to do so is to try to press re-presentation or translation through the grid of language. But this has a speciesist bias in favor of humans. Cats, catflaps, computers, and fax machines — not to mention X-ray sources, safety interlocks, and keys — *all* these order and organize, create paths and links. All of these signify, but they do so in their own ways.

So language isn't the only way of producing signification. There are nondiscursive ways too. Or ways that are only partially discursive. Which means that re-presentation in language may be different and distinctive. But which also suggests that re-presentations, chains of translation in other material forms, are best not treated as second-class citizens.

The argument, then, is that *signification is more general than meaning*. It is not reducible to *verstehende*, to the language games played by human beings. Machines, persons, texts, *all* are opaque. All are complex. All resist, or have the potential to resist, re-presentation. Like the people and the interlock in the story about the X rays. They have their specificities. They cannot necessarily be reduced to language.

This text tells a story about continuity. Agents, it says, are not different in kind. Or better, their difference is not given in the order of things. If they *appear* to be different — if they appear to embody some of the great dualisms of Western social thought, such as the divisions between mind and body, human and non-human, or human and machine — then this is an outcome, an effect, a product. And it could be otherwise. This is why we have talked about hybrid *collectifs*. For, or so we've suggested, agents are effects generated in configurations of different materials. Which also, however, take the form of *attributions*. Attributions which localize agency as singularity — usually singularity in the form of human bodies.

Attributions which endow one part of a configuration with the status of prime mover. Attributions which efface the other entities and relations in the *collectif*, or consign these to a supporting and infrastructural role.

This paper has suggested that the agents we tend to recognize are those which perform *intentions*. And those which use a *language*. Strategic speakers: those are the hybrid *collectifs* which usually come endowed with agency. Which don't have to put up an argument in order to achieve citizenship in the world of social theory. But we've also argued that it doesn't have to be so. That, indeed, it *is* not so. For re-presentation is a matter of translation. But translation does not have to take the form of language. For here is the bias, the logocentric bias which runs everywhere through social theory. The bias in favor of the speakable. Or, we might add (*pace* Jacques Derrida), the writable. Which is why, though we cannot, to be sure, *say* very much about it, we do not wish to link a notion of agency to linguistic re-presentation. For signification — or so we have suggested — is more general than talk. It comes in all kinds of forms. And some, though only some, we can imagine. Others, no doubt, we will never know. Which means that there are multiform kinds of agency: forms of agency that we can't imagine; forms of agency performed in patterns of translation that are foreign to us; forms of agency that are, for instance, nonstrategic, distributed, and decentered.

MICHEL: Which still leaves us with the question of politics.

JOHN: That's where we started off. With the question of politics. With the suggestion that there must be other ways of thinking about agency than as a combination of Nietzsche and liberalism. We've touched on politics from time to time along the way. But, though it's work that still has to be done, I think we've moved one step further toward a kind of postliberalism.

MICHEL: How so?

JOHN: Well, it's like this. I think we've started to shift, albeit only a little, from the liberal panopticism of ANT. To move from that God-like place where the analyst looks down and counts the poor and huddled masses of scallops and machines. I think we've moved from that to a different and more pluralist place. Though that's a bad choice of word. Very bad. For the word is *liberal*. It implies a homogeneous political space. One in which different equivalent entities have rights, can speak (though only in certain ways), and coexist. A "phallogocentric" form of coding, as Donna Haraway puts it.[32] So I think it's something like this: We should be trying to imagine not a single place — but a set of places. A set of somewhat overlapping places which say, in ways that we can only partially imagine, that there are more things in heaven and earth than are dreamed of in a liberal polity. I don't believe that we can know most of those things. From any place in particular. From another part of the forest. Neither do I think that we can follow them systematically through society. Or enumerate them from one spot. To imagine that we can assimilate the Other in any of its forms is hubris. Instead, it seems to me that these Others will ignore us for most of the time. Instead, they will continue, as they always have, to perform their specific forms of agency to one another. And all that we can do is to say that these performances go on. And then to create appropriately monstrous ways of re-presenting them on those rare occasions when our paths happen to cross and we find, for a moment, that we need to interact with them.[33]

NOTES

This version of our paper benefited immeasurably from the discussions at the Surrey conference. We'd like to thank Stella Harding and Robin Woofitt for organizing the meeting and agreeing to let us publish the paper here. In addition, we'd like to thank Steve Brown and Nick Lee for their paper "Otherness and the Actor Network: The Undiscovered Continent." Steve and Nick observe that ANT isn't only Nietzschean, but also liberal. They're right, of course. Perhaps we'd announce the death of ANT if we still deluded ourselves that we are capable of speaking for this monster. We'd also like to thank Annemarie Mol and Stefan Hirschauer, jointly and severally. Severally, we'd like to note that it's been in discussion with Annemarie that we've come to think that ANT, and the network metaphor itself, is unduly monologic and centered and that it has unfortunate liberal implications. Jointly, we'd like to thank them for their paper "Beyond Embrace: Multiple Sexes at Multiple Sites" (see note 25). We've shamelessly stolen the dialogic possibilities of multilinguistic collaboration from their presentation of that paper. Other acknowledgments are due to Bob Cooper, for pressing the Lyotardian rather than the ANT version of heterogeneity, and thus exposing us to Otherness rather than to differences in materiality; to Marilyn Strathern, for pressing the possibility of partial connections; to Rolland Munro, who carefully read and commented on an earlier draft of this paper; to "Andrew" and his fellow managers, for helping John Law create the stories which we use in this paper; and to the British Economic and Social Research Council and its agent (agent?), the Science Policy Support Group, for funding the research in Andrew's lab within its Research Initiative on the Changing Culture of Science.

1   The conference was held at the University of Surrey on 23–24 September 1993. Organized in the sociology department by Stella Harding and Robin Woofitt, it brought together people in most of the trades we've already mentioned, including philosophers, sociologists, experts in artificial intelligence, psychologists, and writers in STS.

2   Steve Brown and Nick Lee, "Otherness and the Actor Network: The Undiscovered Continent," *American Behavioral Scientist* 36 (1994): 772–90.

3 For a brief account of actor-network theory, see John Law, "Notes on the Theory of the Actor-Network: Ordering, Strategy and Heterogeneity," *Systems Practice* 5 (1992): 379–93.

4 See Vicky Singleton, "Science, Women and Ambivalence: An Actor-Network Analysis of the Cervical Screening Programme" (Ph.D. dissertation, University of Lancaster, 1992); and Leigh Star, "Power, Technologies and the Phenomenology of Conventions: On Being Allergic to Onions," in *A Sociology of Monsters? Essays on Power, Technology and Domination*, Sociological Review Monograph 38, ed. John Law (London, 1991), 26–56.

5 See John Law's attempts to wrestle with the problem, though, in his *Organizing Modernity* (Oxford, 1994).

6 The laboratory in question is Daresbury SERC Laboratory, near Warrington in Cheshire, England. For a fuller discussion of the ethnography, see Law, *Organizing Modernity*.

7 The reference is to Michel Callon and Bruno Latour, "Unscrewing the Big Leviathan," in *Advances in Social Theory and Methodology*, ed. Karin D. Knorr-Cetina and Aaron V. Cicourel (Boston, 1981), 277–303. The article considers the materials of sociality available to the (supposedly) dominant male baboon as he seeks to secure and maintain his position — a very small repertoire compared with those available in human society.

8 Paul Auster, *Moon Palace* (London, 1990).

9 Though, no doubt, one that has its limits. See Annemarie Mol and John Law, "Regions, Networks and Fluids: Topology and Anaemia," *Social Studies of Science* 24 (1994): 641–71.

10 See his account of what he calls "the civilising process," in Norbert Elias, *The History of Manners* (Oxford, 1978). And he's not alone. See, for instance, Richard Sennett, *The Fall of Public Man* (London, 1986); and other radical writers, such as Donna Haraway, "A Manifesto for Cyborgs," in *Feminism/Postmodernism*, ed. Linda J. Nicholson (New York, 1990), 190–233.

11 For discussion of organizing and organization theory, see Robert Cooper and John Law, "Organization: Distal and Proximal Views," in *Research in the Sociology of Organizations* 14, ed. S. Bacharach and P. Gagliardi (Greenwich, CT, 1995).

12 See, for instance, Andrew Pickering and Adam Stephanides, "Constructing Quaternions: On the Analysis of Conceptual Practice," in *Science as Practice and Culture*, ed. Andrew Pickering (Chicago, 1992), 139–67.

13 See, for instance, Norbert Elias, *The Court Society* (Oxford, 1983); and Fredric Jameson, *Postmodernism, or, The Cultural Logic of Late Capitalism* (Durham, NC, and London, 1991).

14 For discussion, see Malcolm Ashmore, "Behaviour Modification of a Catflap: A Contribution to the Sociology of Things," *Kennis en Methode* 17 (1993): 214–29.

15 This is reflexivity in the high-modernist sense intended by such authors as Anthony Giddens, *The Consequences of Modernity* (Cambridge, 1990); and Zygmunt Bauman, *Modernity and the Holocaust* (Cambridge, 1989), rather than the reflexivity explored by writers such as Malcolm Ashmore, *The Reflexive Thesis: Wrighting Sociology of Scientific Knowledge* (Chicago, 1989).

16 The concern with language use is central to most forms of interpretive and *verstehende* sociology, English language discourse analysis, and ethnomethodology. These

often trace their roots back to Ludwig Wittgenstein, *Philosophical Investigations* (Oxford, 1953). Examples in the present context include Ashmore, "Behaviour Modification."

17   For this argument, see Bruno Latour, *We Have Never Been Modern,* trans. Catherine Porter (Cambridge, MA, 1993); and Michel Serres, *Eclaircissements* (Paris, 1992).

18   For examples of sophisticated historicisms, see David Harvey, *The Condition of Postmodernity: An Enquiry into the Origins of Cultural Change* (Oxford, 1990); and Jameson, *Postmodernism.*

19   There's a family of roughly related notions here. Michel Foucault, *Discipline and Punish* (Harmondsworth, 1979), comes first with his account of Bentham's panopticon. For the notion of a center of translation, see Bruno Latour, *Science in Action* (Cambridge, MA, 1987); Michel Callon and John Law, "On the Construction of Sociotechnical Networks," *Knowledge and Society* 8 (1989): 57–83; and Bruno Latour, "Drawing Things Together," in *Representation in Scientific Practice,* ed. Michael Lynch and Steve Woolgar (Cambridge, 1990), 19–68.

20   For further discussion of discretion, agency, and the mode of ordering, see Law, *Organizing Modernity.*

21   For discussion of the relational character of materials, see John Law and Annemarie Mol, "Notes on Materiality and Sociality," *Sociological Review* 43 (1995): 274–94.

22   See Monica Casper, "The Social Construction of 'Human' in Experimental Fetal Surgery: Reframing the Debate around 'Non-Human' Agency" (paper presented at the conference "Non-Human Agency: A Contradiction in Terms?" University of Surrey, 1993).

23   This can be extracted from all sorts of traditions. The particular formulation here is drawn from Anthony Giddens, *The Constitution of Society* (Cambridge, 1984).

24   Essentially the same idea is developed in Barry Barnes, *The Nature of Power* (Cambridge, 1988); and Michel Callon, "Socio-Technical Networks and Irreversibility," in Law, ed., *A Sociology of Monsters?,* 132–61.

25   For some of these possibilities in the area of gender, see Haraway, "Manifesto for Cyborgs"; and Stefan Hirschauer and Annemarie Mol, "Shifting Sexes, Moving Stories: Feminist/Constructivist Dialogues," *Science, Technology and Human Values* 20 (in press). For a recent collection which explores not only feminist diversity, but also that of class and ethnicity, see *Place and the Politics of Identity,* ed. Michael Keith and Steve Pile (London, 1993); see also Marilyn Strathern, *Partial Connections* (Savage, MD, 1991), for a beautiful study of partiality.

26   Consider, for example, the conversations between children about the possibility of computer agency described by Sherry Turkle, *The Second Self: Computers and the Human Spirit* (New York, 1984); and the modes of ordering explored in Law, *Organizing Modernity,* where it is assumed, without further reflection, that agency (whether human or non-human) is attributable to singularities rather than to dispersions.

27   See Sennett, *Fall of Public Man.*

28   For collections which explore this debate, see Nicholson, ed., *Feminism/Postmodernism;* and Keith and Pile, eds., *Place and the Politics of Identity.* In social studies of science, see Haraway, "Manifesto for Cyborgs"; Star, "Power, Technologies"; and Singleton, "Science, Women and Ambivalence." For an entertaining, if contentious, argument that the Japanese were always postmodern and, accordingly, do not have to decenter themselves, see John Clammer, "Aesthetics of the Self: Shopping and

Social Being in Contemporary Urban Japan," in *Lifestyle Shopping: The Subject of Consumption*, ed. Rob Shields (London, 1992), 195–215.

29   See, for instance, Umberto Eco, *Lector in Fabula: Le Role du Lecteur* (Paris, 1985).

30   See chapter 4 of Jameson, *Postmodernism*.

31   See Ashmore, "Behaviour Modification."

32   See Haraway, "Manifesto for Cyborgs."

33   On the role of the monster, see John Law, "Introduction: Monsters, Machines and Sociotechnical Relations," in Law, ed., *A Sociology of Monsters?*, 1–23.

The Accidental Chordate: Contingency

in Developmental Systems

Susan Oyama

T HE somewhat peculiar title of this essay is not just an allusion to items of recent popular culture. It refers as well to Stephen Jay Gould's book *Wonderful Life,* in which he discusses the fossils of the Burgess Shale, a rich bed of paleontological remains in Canada.[1] Many of these fossils belonged to extinct phyla. An extinct species or genus is hardly remarkable to most people, but the disappearance of groups as high in the evolutionary hierarchy as the phyla is perhaps more arresting: consider that the vertebrates, as varied as they are, comprise only one subdivision of the phylum Chordata.

Gould enlists the bizarre Burgess creatures, which he calls "weird wonders," in the service of one of his favorite themes, in both his popular and his scholarly writing: *contingency* in evolution. Like many other scientists, Gould frequently argues against the widespread idea that the course of evolution is somehow *necessary:* progressive, goal-directed, always moving from the less to the more complex, culminating in those marvels of reflective intelligence, ourselves. Again like many others, however, Gould has no difficulty attributing these qualities of orderly goal direction to *development,* hence his use of the common metaphor of the computer program.[2]

The two words "evolution" and "development," in fact, have intertwined histories that reflect, at least in part, science's changing understandings of what we would now call the evolution of populations and the development of organisms. It is the latter that has been the primary subject of my own work; more precisely, I have been interested in the various faces and functions of nature/nurture oppositions (innate/acquired, biological/cultural, physical/mental, etc.) and thus have spent considerable time thinking about the way development is conceptualized. Insofar as the "nature" side of the nature/nurture contrast is supposed to be the legacy of our evolutionary past, and insofar as evolution entails changes in patterns of development, I have necessarily pondered the relationship *between* these two formational processes as well.

Gould is interested in developmental constraints on natural selection. Some intriguing linkages exist between the constraints/selection dichotomy and the one between nature and nurture, but I do not discuss them here.[3] Nor do I evaluate the problems of extinction that are presented in *Wonderful Life*. The discussion of the Burgess Shale creatures serves, rather, as a pretext for examining the above-mentioned use of contingency to *contrast* evolutionary and developmental processes. I argue for a notion of development in which contingency is central and constitutive, not merely secondary elaboration of more fundamental, "programmed" forms. Gould uses contingency in two major ways, as unpredictability and as a certain kind of causal dependency. It is useful for my own purposes to distinguish the epistemological from the ontological sense of the word in order to examine the usual assumption that while chordates, or any other phylum, may be *evolutionarily* contingent, any particular chordate (or, for that matter, any organism) is hardly contingent *ontogenetically*, but instead is brought into being by an internal program or plan. My discussion of contingency in developmental processes, which concludes with some thoughts on contingency in theorizing about developmental systems, also notes some possible connections to recent thinking in critical theory and the sociology of science. These connections startled me when I first confronted them, and, as a relative newcomer to these literatures, I offer them tentatively. But perhaps they are not so surprising after all: insofar as the modernist project consists of the attempt to separate nature from culture, a serious rethinking of biological nature and environmental nurture would seem very much to the point.[4]

Gould's argument in *Wonderful Life* is that the very existence of the phylum Chordata, and therefore ourselves, was an evolutionary accident. He repeatedly uses the metaphor of the lottery and says, "Hence, this model strongly promotes the role of contingency, viewed primarily as unpredictability, in evolution."[5] It could seem, then, that his primary concern is epistemological. This impression is strengthened when, in the same passage, he challenges his readers to contemplate the elaborate designs of these extinct creatures and to say what defect explains why they, and not others, should have vanished from the face of the earth. But simple prediction (or retrodiction) is surely not the only point. Gould is also making an argument about what kinds of causal *processes* are involved in evolution, "an unpredictable sequence of antecedent states, where any major change in any step of the sequence would have altered the final result. This final result is therefore dependent, or contingent, upon everything that came before — the unerasable and determining signature of history."[6]

Much could be said about this passage, but for now we may note that Gould is joining the epistemological issue of *predictability* with the ontological one of the nature of evolutionary *process*, as chains of causal dependency of a certain sort. Gould certainly didn't invent the association between these particular meanings of contingency.[7] Nor is the association always pernicious. As noted above, however, the distinction between the predictability of processes and the nature of those processes will be crucial to my discussion of developmental dynamics.

Development involves the repeated arising and transformation of complexes of interacting processes and entities. These occur not because of some preordained necessity and not, obviously, by "mere chance," if this means an absence of regularity or causal relation, but by systems of contingencies whose organization may itself be contingent and, in any case, must be accounted for. Reliable, species-typical life courses can be seen as contingent in a number of ways (not absolutely necessary, causally dependent on factors that may in turn be uncertain) while still being highly predictable, and thus noncontingent in another sense. Unlike evolution, development presents us with repeating cycles. Despite many kinds of variation, these can be so similar across generations that they offer themselves for comparison with the transmission of property in accordance with social institutions of inheritance. They thus invite confusion between the epistemological and the ontological aspects of contingency.

The narrowing of the life cycles of some organisms to a single cell, and the identification of fundamental biological form and function with the DNA that passes through this "developmental bottleneck," has encouraged the practice of endowing genes, which provide one sort of continuity between generations, with the capacity to generate, virtually de novo, the next cycle. In fact, the lesson usually taken from the existence of these bottlenecks is that the DNA (or, slightly more generously, the germ cell) must control the development of the next generation. The sheer *predictability* of many aspects of development has been explained by invoking special causal *processes:* centrally controlled maturation, predetermined by coded "information" or "instructions" transmitted in molecular agencies that are immortal, omnipotent, omniscient, and even immaterial. This involves a systematic privileging of insides over outsides and active, controlling agents over passive materials: the environment is seen as supporting, modulating, and constraining development, while primary formative power is reserved for the genes.[8]

An alternative explanatory model is found in the developmental systems approach, which explains such reliably repeated sequences of events by invoking heterogeneous, complexly interacting, and mutually constraining entities and processes in which "control" is distributed and fluid rather than centralized and fixed.[9] The successive reconstruction of life cycles is possible not because genetic programs or controlling agents are transmitted and safely insulated from the outside world by the nuclear boundary, but precisely because what is inside the cell interacts with what is outside it. Cellular processes occur by means of interactions among microscopic constituents, and, on larger scales, organisms interact with their worlds. The necessary resources are reconstituted (e.g., DNA by replication, cells by division), provided by maternal physiology and behavior, supplied or sought by the organism itself and by conspecifics or by other organisms, including symbionts. They may also be intermittently or continuously present in the nonanimate surround.[10]

Whether or not an element or variety of energy is a "resource" depends on its relation to the developing organism, which is in turn defined and constructed by its internal and external interactions. In this view, a gene is a resource among others rather than a directing intelligence using resources for its own ends. There is, finally, no centralized repository of "information" and causal potency that explains the repeated lives of organisms, no matter how much our notions of biological necessity may seem to require one. There are, however, many ways in which processes that have usually been considered independent of each other can be seen as actually impinging upon each other; these relations are part of the developmental story.

If one adopts this way of approaching vital processes, many vexatious distinctions become not only unnecessary, but unintelligible (features that are inherited vs. those that must develop, those shaped by genes vs. those shaped by learning or "the environment," the biological vs. the cultural, at least when these distinctions imply differing formative roles of genes and environment), while reasonable distinctions can be unambiguously stated and so treated with less confusion. One can compare, that is, features that vary within a species with those that are species-typical, outcomes that seem difficult to perturb under a certain range of circumstances with those that vary in that same range, behavior that is present at birth with behavior that develops later, and so on. The point is that these are different *questions*, not different ways of approaching the same question.

The notion of repeated cascades of contingencies, some more tightly constrained than others, has been central to work on developmental systems. Such a system is composed of all developmentally relevant influences interacting over the life cycle to produce, maintain, and alter the organism and its changing worlds. I call these influences, whether animate or inanimate, "interactants." Although this term was formulated independently of Bruno Latour's and Michel Callon's work on actants in techno-economic networks, there are some striking conceptual similarities, which I believe to be nonaccidental (albeit unintended).[11] A developmental system is emphatically not bounded by the organism's skin, but includes nested systems on a variety of scales from the molecular to the ecological.

The results of these developmental processes cannot be attributed to the formative power of one class of factors whose action is constrained or modulated by other, secondary factors, as traditional schemes require, but rather are the (moving) outcomes of interactions in the system. Interactants and processes change over ontogenetic and phylogenetic time. Some are more reliable than others: The term "system" should not be taken as a guarantee of absolutely faithful replication (or of unfailing regulation in the embryological sense of restoration of a normal outcome despite perturbation), but rather as a marker of a complex, interacting network that may, and often does, arrange its own relatively accurate repetition. "System" implies self-organization, where "self" is not some privileged constituent or prime mover, but an entity-and-its-world, which world is extended and heterogeneous, with indeterminate and shifting boundaries.[12] The developmentally relevant world or environment here is not organism-independent, but is interdefined with it; since both the organism and its environment are changing, partly as a result of their interactions, the system's constituents and processes also change over time. Evolution, then, is change in the constitution or distribution of these systems. For those who insist on the neo-Darwinian synthesis definition of evolution in terms of allele (variant gene form) frequencies, it is still possible to bring only these elements to the fore. To do so requires a quite arbitrary privileging of the molecular level (in fact, only one aspect of it), however, and the gene-centered view has a variety of regrettable consequences, including questionable notions of developmental control. The whole thrust of developmental-systems thinking is to reject, rather than to endorse, this kind of privileging.

In speaking of unpredictable historical sequences, Gould says that "contingency precludes [their] repetition, even from an identical starting point."[13] But developmental contingencies, and at least similar starting points, are what *allow* repetition, if and when it occurs. Causal dependence on uncertain conditions (one definition of "contingency") needn't involve unique, unrepeatable, and, in principle, unpredictable sequences. In the repeating life cycles discussed above, the conditions for various interactions are dependable to varying degrees. One needs to know when and how those conditions become more or less certain (How reliable are the formational processes?) and how crucial it is that precisely those conditions be present (How forgiving are they?), questions that are themselves related. Because formational robustness is dependent on other factors, one should not really speak of "canalized" or "buffered" characters as though this were somehow inherent in the character itself. Human limb formation is reliable under many but certainly not all gestational conditions; think of thalidomide. To continue the chain of conditionals, such conditions have different effects on different people, in different settings, at different times. (Thalidomide in the eighth gestational month has different effects from those caused by the drug in the third month, and there are surely individual differences in responsiveness.)

Consider this example of an assembly of contingencies (Figure 1). In this invention by cartoonist Rube Goldberg, the adult desirous of sleep must initiate the following sequence:

> Pull string (A) which discharges pistol (B) and bullet (C) hits switch on electric stove (D), warming pot of milk (E). Vapor from milk melts candle (F) which drips on handle of pot causing it to upset and spill milk down trough (G) and into can (H). Weight bears down on lever (I) pulling string (J), which brings nursing nipple (K) within baby's reach. In the meantime baby's yelling has awakened two pet crows (L & M) and they discover rubber worm (N) which they proceed to eat. Unable to masticate it, they pull it back and forth causing cradle to rock and put baby to sleep.[14]

Here's another example (Figure 2). Natalie Angier, of the *New York Times*, says of recent discoveries about the *ras* pathway in cell division

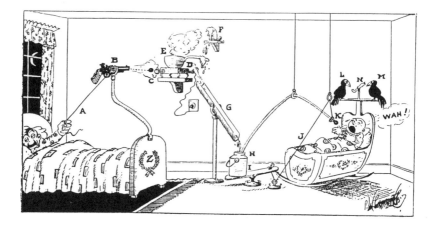

Figure 1. "Inventions of Professor Lucifer Butts: Anti-Floorwalking Paraphernalia" by Rube Goldberg. *Cartoon Cavalcade,* ed. Thomas Craven (New York: Simon and Schuster, 1943), 235.

(the *ras* gene, which gets its name from *rat* sarcoma, is believed to be implicated in many cancers):

> Molecular biologists are now gazing upon their glittering prize, a fundamental revelation into how the body grows. The sight is astonishing to behold. It is epic.
>
> It deserves a crash of cymbals, a roll of the tympanum — and a hearty guffaw. It turns out the much-exalted signaling pathway of the cell is a kind of molecular comedy, in which one protein hooks up to a second protein that then jointly push a button on an enzyme that pushes a button on another enzyme that makes this knob slide into that hole — all in all like something Wile E. Coyote might have pieced together from one of his Acme kits.[15]

The joke here has nothing to do with unrepeatability: Angier continues, "Despite its improbability, its cartoonish complexity, the design works wondrously in overseeing cell growth, so well that it is shared by species as distantly related as yeast, worms, flies and humans." Rather, the comedy seems to derive from the violation of our notions of simplicity, of logical necessity. One scientist involved in *ras* research asks plaintively: "Why is it all so complicated? Why do you need so many steps?"[16] Biologist Sydney Brenner made similar comments on his failing to find the developmental program for the tiny, much-studied roundworm *Caenor-*

habditis elegans. Brenner complained that cell lines were "baroque" and that there seemed no briefer way of describing what happened than simply giving an account of the entire sequence of events.[17] (Notice the unspoken assumption of informational compressibility, of the possibility of giving a rule that is shorter than a "mere description" of what happens.[18])

In the *ras* pathway, and in the ontogeny of Brenner's worm, sequences are "improbable" in that they would have been hard to predict before the fact. They have an arbitrary, uneconomical air that offends the sensibilities of those seeking the spare elegance of "law"; each event seems to occur not because of some transcendental necessity, but because the constituents just happen to be lying around. The whole point, of course, is that they *do* lie around, enabled and constrained by a precise spatiotemporal organization that becomes possible only in a richly differentiated setting. Consider, for example, the enforced propinquity of diverse molecules in a highly organized cell. These objects, relations, and reactions are capable of assembling, if temporarily, into systems of

Figure 2. A segment of the *ras* pathway, thought to be involved in the development of cancers in a variety of species. Illustration by Baden Copeland. *New York Times*, 29 June 1993, C10. Reproduced with permission.

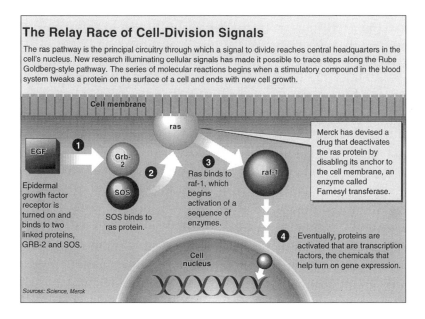

local necessities: contingently predictable congeries of dependencies and interdependencies, recurring cascades of contingencies.

The *ras* narrative takes place, for the most part, in the interior of individual cells, but these linked cascades occur on larger scales as well and can include stable social processes, which in turn can feed back into physiological events. As long as "environmental" effects are understood as accidental, however, they are unlikely to be integrated into accounts of ontogenesis. Goldberg's "anti-floorwalking" devices must be reset for their next use (though he also recommends that the adult keep earplugs handy in case the baby awakes again). Part of the functioning of many biological processes, however, is the entrainment of the next cycle. As noted above, this becomes possible precisely because the processes are not isolated, insulated, or autonomous, but connected to others, by which they are influenced, and which they may in turn influence. (Because these enabling/constraining systems do not stop at the border of the skin, the developmental-systems perspective requires a willingness to cross familiar boundaries in tracing such connections. This flexibility allows seasonal or other ecological regularities, for example, to be included in the developmental account.[19])

In a discussion of the history of life, Jack Cohen tells of the appearance of self-feeding chemical reactions, where "each reaction affected some other process which led through a series of steps back to control of itself." What was important about such feedbacks, he notes, was that they could "*keep the set-up going,* instead of hastening it to some kind of completion or exhaustion."[20] These multiple dependencies ultimately make the metaphor of the linear chain inapt, although a scientist may excise part of the process to analyze as if it were an isolated chain running off autonomously against the background of the rest of the system. To do so, however, all of that background must be held constant (treated as *given* as well as kept from varying), just as Goldberg stabilizes by fiat, and is drolly mute about, many of the connections and contingencies that bring together, and hold together, his precarious assemblies.

Goldberg's contraptions also make ingenious use of just the kind of human/nonhuman/machine hybridization spoken of by Latour and Callon in their respective analyses of networks.[21] Notice that the desires and actions of the crows are assumed to be as stable as the melting point of paraffin. In another of Professor Butts's inventions, a fan's components include not only a wheelbarrow and a doll, but a live lovebird that nods every time it is asked whether it loves a mechanical bird, and a bear which, when "annoyed" at being kicked, "suspects" the doll dangling before it and eats it, thus triggering the next step in the astonishing concatenation.[22]

We have seen that the epistemological question of predictability is not always distinguished from the ontological one of process in discussions of contingency, that reliably repeated assemblies can be noncontingent in the sense of being highly predictable, while being thoroughly contingent in their dependence on complex, elaborately extended systems of interacting factors whose dynamic organization cannot be explained in terms of a single component or central agency. I conclude by suggesting that we extend the habit of remembering contingency, of saying, "It depends . . . ," to the activity of theorizing itself.

Characteristic of much of the work on developmental systems is an insistence on a sort of parity of reasoning. If, for example, some aspect of an organism is deemed to have a "biological base" because its variants are correlated with genetic variation in a particular population at a particular time, then one ought to be willing to call something "environmentally based" if it is correlated with variations in the surround. As pointed out by many commentators on the nature/nurture opposition, the same feature may thus be "wholly biological" and "wholly environmental," depending on conditions and the investigative choice of the researcher. Or, if some bit of matter is termed a "master molecule" because certain interesting things happen after it is activated (this is a loose inference about usage; the reasoning behind such terms is seldom explicit), then the same criteria should be used for all links in the sequence having similarly interesting sequelae — and, of course, what is interesting, and to whom, is hardly fixed or uniform.[23] Obviously, the point of such exercises in conceptual equity is not to distribute causal honorifics to more and more entities, but rather to question them by rendering them explicitly situation-specific. By undermining the raison d'être of such terms, which is to elevate certain elements above others, the practice of relativization-by-contextualization should curb the impulse toward wildly extending the range of application.

One virtue of the parity-of-reasoning arguments, which in many ways resemble the "symmetry" rule described by Callon and by Barbara Herrnstein Smith, for example, is that they direct attention to the often-unrecognized assumptions informing the questions we ask and the ways we ask them.[24] To emphasize the relativity of questions to the assumptions and purposes that underlie them is not to claim that all entities, questions, or assumptions are equal or that it doesn't matter what one

thinks because all ideas are equally valid.[25] On the contrary, precisely because it *matters very much* what questions we ask and why, and because
the meaning of a question, as well as its possible answers, depends on
what kind of world it takes for granted, it is important to ask on what
vision of the world those questions depend. Even the boundaries of a
developmental system are relative to the type of inquiry one is conducting, to the kind of story one wants to tell. A social-psychological account
may include elements that an evolutionary one ignores.

Consider a scar, perhaps one incurred in a duel. Paul Griffiths and
Russell Gray would exclude a scar on the hand from the developmental system because it would not contribute to its own reconstruction
in the next generation (as a heart, for instance, would).[26] But a scar
carrying a certain symbolic/social weight might do so, at least in a short-
term and general way, under certain social circumstances. A saber duel
might give immediate resolution to a dispute, but also have a medium-
term tendency to stabilize a person and family in the social system
(with who knows what consequences for their reproductive success!)
and a longer-term tendency to legitimize and perpetuate the social system itself, including its customary methods of resolving disputes, and
thus the occurrence and the meaning of certain kinds of scars. In any
case, I would distinguish between what might be termed "evolutionary
developmental systems" and those with features that may not be transgenerationally stable, but may play an important role in a particular
life nonetheless (and so might be of interest to a developmental psy-
chologist, say, or a sociologist). As suggested above, I would also permit
gradations between the two kinds of systems. Such gradations are not
only convenient, but theoretically crucial for showing the continuity between cultural and biological evolution (a goal shared by Griffiths and
Gray).[27] Dueling scars may not recur across hundreds of generations,
but circumcisions, for example, can be transgenerationally stable for
long periods and important in maintaining both cultural continuity and
non-random reproduction. (Perhaps the ultimate case of such a developmental "accident" is the navel, and it would be a nice exercise to test
the various criteria for nature and nurture on it.)

Showing that many ways of privileging the gene turn out to be quite
unjustified can result in the inclusion of more kinds of phenomena in
an initially restricted scheme, as levels of selection or even replicators
proliferate when a particular logic is applied to more and more cases.[28]
But, as suggested in my earlier comments on master molecules, it can
also call into question the very language of self-replication, information,
autonomous control—language that may become untenable when it is

properly contextualized. Consider, again, Angier's description of just some of the contingencies involved in the *ras* protein's workings and her willingness to refer to just this molecule as "the mastermind of a signaling cascade." One could also, I suppose, call the rubber worm in Rube Goldberg's machine the mastermind of baby-soothing.

Although Angier's account appeared in the *New York Times,* such language is frequently found in the scientific literature as well. More to the point, it is tied to a way of thinking about centrally controlled biological processes that can be quite problematical. As has often been noted, such oppositions as autonomy versus connection, central versus distributed control, and nucleus versus cytoplasm have rich metaphorical associations that are neither scientifically nor socially inert.[29] Latour has described the scientist as a spokesperson for that which is studied;[30] I suppose that one of the many reasons I have found it worthwhile to think and write in developmental-systems terms is that it allows me to speak for the background—the mute, manipulated materials, the featureless surround. Sometimes, the peripheral is the political.

Ultimately, it may seem less appropriate to speak of entities' replicating themselves (thus marginalizing and instrumentalizing everything else) than to say that they may be assembled or constructed again and again as a result of processes in which they play a part but do not "control," except when one isolates a part of the process for analysis.

Seeing erstwhile prime movers as players in a game they "control" only according to a particular framing of a question and only by being "controlled" by other factors doesn't keep us from making distinctions, from studying processes or intervening in them. It may, however, make us less likely to underestimate the constitutive importance of these "other factors," and thus keep us from being quite so surprised when it all turns out to be so complicated, or when unintended consequences and overlooked (but necessary, after all) conditions (or people) force us to look again, for these are just what are pushed into the dim background by more monomaniacal stories.[31]

By emphasizing the contingency of humans (introducing a distinctly unprepossessing Burgess chordate at the very end of his book), Gould denies predestination and draws a moral lesson about the importance of choice and action. But if theorizing about contingency is itself contin-gent in the ways I have suggested, we can turn Gould's moral around: It is equally important to recognize the "choices" we have *already* made, however unreflectively or tacitly. Indeed, it is essential to articulate them and own them, or even, on second thought, once we have looked at them closely and related them to our other beliefs and concerns, to put them aside and make better ones. Taking some factor for granted or including it in a ceteris paribus clause doesn't mean that it plays no formative role or that it will always be there, something we realize with growing alarm as developmental, social, and ecological systems go awry, forcing closer attention to those "background" conditions that account for both the robustness and the vulnerability of developmental systems.

NOTES

This essay is based on a talk delivered at the July 1993 meeting of the International Society for the History, Philosophy, and Social Studies of Biology, at Brandeis University. The symposium at which it was given was organized by Linnda Caporael and Elihu Gerson and was originally entitled "Contingency" (later retitled "Repeated Assembly").

1   Stephen Jay Gould, *Wonderful Life: The Burgess Shale and the Nature of History* (New York, 1989).

2   See Stephen Jay Gould, *Ontogeny and Phylogeny* (Cambridge, MA, 1977), 18. For discussion of the computer program metaphor, which is a commonplace of biology and other fields, see Brian C. Goodwin, "Biological Stability," in *Towards a Theoretical Biology*, ed. C. H. Waddington (Chicago, 1970), 3: 1–17; H. F. Nijhout, "Metaphors and the Role of Genes in Development," *BioEssays* 12 (1990): 441–46; and Susan Oyama, *The Ontogeny of Information* (Cambridge, 1985), chapter 5.

     Although the issue of progress or directionality in evolution is not closed, it tends to be discussed by evolutionists and philosophers in rather more rarified terms; see *Evolutionary Progress*, ed. Matthew H. Nitecki (Chicago, 1989). The emphasis on (nonadaptive) chance in *Wonderful Life* is part of a much larger argument about hierarchy in evolution and the adequacy of natural selection as an explanation for all levels of evolutionary change. For an outline of this argument, see Stephen Jay Gould, "Is a New and General Theory of Evolution Emerging?" in *Evolution Now*, ed. John Maynard Smith (San Francisco, 1982), 129–45.

3   See Russell D. Gray, "Beyond Labels and Binary Oppositions: What Can Be Learnt from the Nature/Nurture Dispute?" *Rivista di Biologia/Biological Forum* 80 (1987): 192–96; and Susan Oyama, "Ontogeny and Phylogeny: A Case of Metarecapitulation?" in *Trees of Life: Essays in Philosophy of Biology*, ed. Paul E. Griffiths (Dordrecht, 1992), 211–39.

4   On this view of the modernist project, see Bruno Latour, *We Have Never Been Modern*, trans. Catherine Porter (Cambridge, MA, 1993).

5   Gould, *Wonderful Life*, 308.

6   Ibid., 283–84.

7   See S. M. Cahn, "Chance," in *Encyclopedia of Philosophy*, ed. Paul Edwards (New York, 1967), 2: 73–75. See also W. H. Dray, on "Determinism in History," 373–78, in the same volume. Both ontogeny and phylogeny, of course, are historical phenomena, and it is precisely the conceptualization of history that is at issue here.

8   On bottlenecks, see John T. Bonner, *On Development* (Cambridge, MA, 1974). For influential characterizations of genes, see Richard Dawkins, *The Selfish Gene* (Oxford,

1974); or George C. Williams, *Adaptation and Natural Selection* (Princeton, 1966). For comment on George C. Williams's description of genes as immaterial packets of information, see Susan Oyama, "Stasis, Development and Heredity," in *Evolutionary Processes and Metaphors*, ed. Mae-Wan Ho and Sidney W. Fox (New York, 1988), 255–74. Analysis of relevant metaphors can be found in Richard M. Doyle, "On Beyond Living: Rhetorics of Vitality and Post Vitality in Molecular Biology" (Ph.D. diss., University of California at Berkeley, 1991); Evelyn Fox Keller, *Reflections on Gender and Science* (New Haven, 1985); and Oyama, *Ontogeny of Information*, especially chapter 5.

9   See, for example, Russell D. Gray, "Death of the Gene: Developmental Systems Strike Back," in Griffiths, ed., *Trees of Life*, 165–209; Paul E. Griffiths and Russell D. Gray, "Developmental Systems and Evolutionary Explanation," *Journal of Philosophy* 91 (1994): 277–304; Timothy D. Johnston and Gilbert Gottlieb, "Neophenogenesis: A Developmental Theory of Phenotypic Evolution," *Journal of Theoretical Biology* 147 (1990): 471–95; Susan Oyama, "A Reformulation of the Concept of Maturation," in *Perspectives in Ethology*, ed. P. P. G. Bateson and P. H. Klopfer (New York, 1982), 5: 101–31; and Oyama, *Ontogeny of Information*.

Two important earlier works in a similar spirit are Daniel S. Lehrman, "Semantic and Conceptual Issues in the Nature-Nurture Problem," in *Development and Evolution of Behavior*, ed. Lester R. Aronson, Ethel Tobach, Daniel S. Lehrman, and Jay S. Rosenblatt (San Francisco, 1970), 17–52; and Richard C. Lewontin, "Organism and Environment," in *Learning, Development, and Culture*, ed. Harold C. Plotkin (New York, 1982), 151–70. There are also many points of contact between the developmental-systems approach and that presented in Humberto R. Maturana and Francisco J. Varela, *The Tree of Knowledge: The Biological Roots of Human Understanding* (Boston, 1988); and Francisco J. Varela, Evan Thompson, and Eleanor Rosch, *The Embodied Mind: Cognitive Science and Human Experience* (Cambridge, MA, 1991).

10   See Jack Cohen, "Maternal Constraints on Development," in *Maternal Effects in Development*, ed. D. R. Newth and M. Balls (Cambridge, 1979), 1–28; Gilbert Gottlieb, "Conceptions of Prenatal Development: Behavioral Embryology," *Psychological Review* 83 (1976): 215–34; Meredith West and Andrew King, "Female Visual Displays Affect the Development of Male Song in the Cowbird," *Nature* 334 (1988): 244–46; and Lynn Margulis, *Symbiosis in Cell Evolution* (San Francisco, 1981).

11   See Bruno Latour, "Technology Is Society Made Durable," in *A Sociology of Monsters? Essays on Power, Technology and Domination*, Sociological Review Monograph 38, ed. John Law (London, 1991), 103–31; and Michel Callon, "Some Elements of a Sociology of Translation: Domestication of the Scallops and the Fishermen of St Brieuc Bay," in *Power, Action and Belief: A New Sociology of Knowledge?* Sociological Review Monograph 32, ed. John Law (London, 1986), 196–233.

12   Compare the relations between object and context in Michel Callon, "Techno-Economic Networks and Irreversibility," in Law, ed., *A Sociology of Monsters?* 132–61; and between person and setting in Jean Lave, "The Values of Quantification," in Law, ed., *Power, Action and Belief*, 88–111.

13   Gould, *Wonderful Life*, 278.

14   Rube Goldberg, "Inventions of Professor Lucifer Butts: Anti-Floorwalking Paraphernalia," in *Cartoon Cavalcade*, ed. Thomas Craven (New York, 1943), 235; reprinted from *Colliers Magazine* (1932).

15   Natalie Angier, "Researchers Track Pivotal Pathway That Makes Cells Divide," *New York Times*, 29 June 1993, C1.

16   Ibid., C10. Angier is quoting *ras* investigator Anthony Pawson here.

17   Quoted in Roger Lewin, "Why Is Development So Illogical?" *Science* 224 (1984): 1327–29; quotation on 1327.

18   On description and explanation, see Callon, "Techno-Economic Networks," 154–55.

19   See Griffiths and Gray, "Developmental Systems," for one way of individuating these complexes.

20   Jack Cohen, *The Privileged Ape* (Park Ridge, NJ, 1989), 11; his emphases.

21   See Latour, "Technology"; and Callon, "Elements of a Sociology of Translation," and "Techno-Economic Networks."

22   Rube Goldberg, "Get One of Our Patent Fans and Keep Cool," in Craven, ed., *Cartoon Cavalcade*, 214.

23   On heritability and development, see Timothy D. Johnston, "The Persistence of Dichotomies in the Study of Behavioral Development," *Developmental Review* 7 (1987): 149–82. On master molecules, see Keller, *Reflections*, 154, quoting David Nanney.

24   See Callon, "Elements of a Sociology of Translation"; and Barbara Herrnstein Smith, "Belief and Resistance: A Symmetrical Account," *Critical Inquiry* 18 (Autumn 1991): 125–39.

25   These objections were raised at the conference at which an earlier version of this paper was given, and they are routinely raised in other contexts as well. There is more than a passing similarity between these objections and those answered, in somewhat different ways, by Latour, "Technology"; by Barbara Herrnstein Smith, *Contingencies of Value: Alternative Perspectives for Critical Theory* (Cambridge, MA, 1988); and "The Unquiet Judge: Activism without Objectivism in Law and Politics," *Annals of Scholarship* 9 (1992): 111–33; and by Susan Leigh Star, "Power, Technology and the Phenomenology of Conventions: On Being Allergic to Onions," in Law, ed., *A Sociology of Monsters?* 26–56. The political and moral issues they address are sometimes intimately enmeshed with the very questions of biological essence and naturalness/normality/necessity that are at stake in the developmental biological and psychological literatures. See also Susan Oyama, "How Shall I Name Thee? The Construction of Natural Selves," *Theory and Psychology* 3 (1993): 471–96.

26   Griffiths and Gray, "Developmental Systems," 286.

27   See Susan Oyama, "Transmission and Construction: Levels and the Problem of Heredity," in *Levels of Social Behavior: Evolutionary and Genetic Aspects*, ed. Gary Greenberg and Ethel Tobach (Wichita, 1992), 51–60; "Rethinking Development," in *Handbook of Psychological Anthropology*, ed. Philip Bock (Westport, CT, 1994), 185–96; and "Bodies and Minds: Dualism in Evolutionary Theory," *Journal of Social Issues* 47 (1991): 27–42.

28   A case in point is the suggestion, made by Kim Sterelny, Kelly Smith, and Michael Dickison, that the developmental-systems perspective leads to extending replicator status to nongenetic, even nonliving, constituents of a repeating life cycle, in "The Extended Replicator," *Biology and Philosophy* (in press). The terms derive from the "replicators" and "vehicles" of Dawkins, *Selfish Gene;* renamed "replicators" and "interactors" by David Hull, *Science as a Process* (Chicago, 1988). Such terms are associated with a particular style of explaining evolutionary change as the machinations of "selfish" entities maneuvering to ensure that replicas of themselves will appear in the next generation. See Griffiths and Gray, "Developmental Systems," for a developmental systems–style alternative.

29   The methodological, metaphorical, and wider social aspects of these matters are not

really separable. See Keller, *Reflections,* especially chapter 8, on the "pacemaker"; and *Secrets of Life, Secrets of Death* (New York, 1992); or Jan Sapp, *Beyond the Gene: Cytoplasmic Inheritance and the Struggle for Authority in Genetics* (Oxford, 1987).

30   Bruno Latour, *Science in Action* (Cambridge, MA, 1987), 71–72.

31   See Susan Oyama, "The Conceptualization of Nature: Nature as Design," in *Emergence: The Science of Becoming,* Vol. 2 of *Gaia,* ed. William Irwin Thompson (Hudson, NY, 1991), 171–84; and Star, "Power, Technology and the Phenomenology of Conventions."

Complementarity, Idealization, and the Limits

of Classical Conceptions of Reality

Arkady Plotnitsky

Questa natura si oltre s'ingrada
in numero, che mai non fu loquela
né concetto mortal che tanto vada.
— Dante, *Paradiso*, Canto 29

T HE principle of complementarity, central to Niels Bohr's interpretation of quantum mechanics, both raises and illuminates
the question of idealization — physical, mathematical, or conceptual — in quantum physics, mathematics, and more generally. It may
be asked, first, what mathematics idealizes (or represents) outside the
context of physics, where it is commonly seen as helping to represent,
perhaps in an *idealized* form, physical processes. This question is old,
much older than modern — post-Galilean — physics, whose project is
both, and codependently, experimental and mathematical.[1] It antedates
both Aristotle and Plato, although many answers relating mathematics
to ideas and ideal entities (rather than to material, physical objects) owe
much to them. A paradigmatic form for such answers, defining what
is sometimes called mathematical Platonism, is that what mathematics
idealizes is itself an ideal, or indeed *the* ideal, mathematical or of some
other kind, which is otherwise unattainable (for example, by physics or
by nature [*physis*] itself).

But then, it may be asked, what does physics — both its theory and, as
we have come to realize, its experimental data — idealize, if it does not
directly, or even indirectly, represent nature or matter, as we used to
(and some still do) believe? This question arises especially in quantum
physics, which, I shall argue here, appears fundamentally to disallow
realist interpretations, in contrast to classical physics, which appears at
least to permit, if not to require, them. As an alternative to realism,
physical Platonism is also possible, of course, conceived either by way
of the mathematical formalism involved or by way of formations that

would exceed any mathematical formalism. If such formations are seen as material (for they may be conceived otherwise), Platonism becomes a form of physical, but extra-mathematical, realism. In general, realism and Platonism can become complicit, sometimes, as Jacques Derrida argues, by uncritically reversing each other.

Finally, in view of postclassical mathematics and science, and a post-classical understanding of both, one may ask, what do our concepts of nature or matter itself represent or idealize? Can they, can anything, still represent or idealize something else, something that is not already — or, as we say now, always already — a representation or an idealization? Can one retain such concepts as representation and idealization under these conditions? Can one not retain them? The complexities involved in formulating the concepts of representation and idealization have always been formidable; the complexities entailed by abandoning these concepts may be more formidable still.

It becomes, and, in truth, has always been, difficult and perhaps finally impossible, at the limit, to complete such phrases as "a representation of . . ." or "an idealization of. . . ." Nietzsche, reflecting on Anaxagoras's proto-quantum-mechanical idea of "a free undetermined *nous*, . . . guided by neither cause nor effect," says that in guessing the nature of such processes "the answer is difficult. Heraclitus supplied: a game [*ein Spiel*] . . . the final solution, the ultimate answer, that ever hovered on the lips of the Greeks."[2] As conceived by Nietzsche, if not by Heraclitus himself, this "answer" has far-reaching consequences.

In outlining a field of concepts responsive to this difficulty or impossibility, my focus here will be on the implications of Bohr's interpretation of quantum mechanics, which may be seen as a culmination in the field of physics of the history inaugurated by these and related pre-Socratic ideas, especially Democritus's conception of the world as the play of chance and necessity. This conception is a key reference for both Bohr and Nietzsche, as for most other figures at the postclassical end of this history, especially Deleuze and Derrida. The main rubric I shall use here is that of *radical alterity*, designating that which is inaccessible to any conceivable mode of representation or idealization, even a representation or idealization that proceeds via the idea of inaccessibility itself. For a number of reasons, the term "alterity" is preferable to "exteriority" or "difference," although these and other terms and concepts can and at certain points must be used in naming (or unnaming) this alterity. The theoretical field at issue cannot be governed by any single term or concept or by any fully determinable or decidable conceptual conglomerate. In accord with both Nietzsche's and Derrida's views, the framework offered here suspends the possibility of unique or final terms, concepts, or frameworks, including itself. Indeed, this framework makes impossible a full conceptual closure or enclosure of any kind.[3] This is why it is preferable to speak here of a field rather than a single concept or rubric, even if it is conceived of as what Derrida calls "neither a term nor a concept."[4] The terms "alterity" and "radical," and all other terms used here — or, conceivably, all conceivable terms and concepts — become provisional and potentially inadequate. In particular, while the alterity at issue must be conceived of as *radical* (i.e., irreducible in relation to any conceivable mode of representation) and while it relates to what is inconceivable, unrepresentable, unidealizable, and unnameable, it cannot be understood as an *absolute* alterity. It cannot be thought of as relating to what is (represented or idealized as) absolutely different from or inaccessible to any representation, as are, for example, Kantian things-in-themselves. Absolute alterity — or anything absolute — would not be radical enough, although such concepts may be necessary intermediate steps in the logic of radical alterity. For one thing, Kant's and other concepts of absolute alterity are powerless against — and indeed share the same metaphysical base with — classical concepts of "overcoming" alterity, such as those of Hegelian and post-Hegelian dialectic, from which one must equally depart. This double departure is not unequivocal, not *absolute*. It can, however, be made sufficiently *radical,* as in the works of not only Bohr, but also Nietzsche, Bataille, Derrida, and several others.[5]

Quantum physics was inaugurated in 1900 by Max Planck's discovery of the quantum nature of radiation and his attempt to formulate and understand the radiation law for the so-called black body, whose model is a heated piece of metal with a cavity. I cannot here consider in detail either Planck's law itself (a remarkable achievement in its own right) or certain related questions in physics.[6] The main point is that his attempt to interpret the law led Planck to the discovery of the quantum character of radiation (that is, an irreducible discontinuity involved in all such processes), which conflicts with the classical understanding of radiation. The limit at which this discontinuity appears is defined by the frequency of the radiation of the body under consideration in a given experiment and by a universal constant (of a very small magnitude) h, or Planck's constant — which turned out to be one of the most fundamental constants of all physics. The indivisible *quantum* of radiation in each case is the product of h and the frequency v, $E = hv$. As Abraham Pais concludes,

Even if Planck had stopped after October 19 [1900], he would forever be remembered as the discoverer of the radiation law. It is
a true measure of his greatness that he went further. He wanted
to interpret [his equation]. That made him the discoverer of the
quantum theory.[7]

Pais has commented elsewhere, "Were I to designate just one single
discovery in twentieth-century physics as revolutionary I would un-
hesitatingly nominate Planck's of December 1900," a view that echoes
both Einstein ("This discovery set science a fresh task: that of finding
a new conceptual basis for all physics") and Bohr ("A new epoch was
inaugurated in physical science by Planck's discovery of the quantum
of action").[8] Pais added that while Planck's "reasoning was mad, . . .
his madness has that divine quality that only the greatest transitional
figures can bring to science. It casts Planck, conservative by inclination,
into the role of a reluctant revolutionary."[9]

Einstein remarked, in his subsequent (1909) analysis of Planck's law,

Our current foundations of the radiation theory have to be abandoned. . . . It is my opinion that the next step in the development
of theoretical physics will bring us a theory of light that can be interpreted as a kind of fusion of the wave and the emission [particle]
theory. . . . [The] wave structure and [the] quantum structure . . .
are not to be considered as mutually incompatible. . . . It seems to
follow from the Jeans law that we will have to modify our current
theories, not to abandon them completely.[10]

(The Jeans law and the Wien law, two disparate laws of black-body radiation operative in different regimes, were introduced prior to Planck's
law, which bridged them.) As Pais observes, "This fusion now goes by
the name of complementarity. The reference to the Jeans law we would
now call an application of the correspondence principle."[11]

Both complementarity and the correspondence principle — which
states that within classical limits the results of classical and quantum
theories coincide, and which, as Bohr put it, "expresses our endeavour
to utilize all the classical [physical] concepts, giving them a suitable
quantum-theoretical re-interpretation"[12] — are among Bohr's own major
contributions to physics. The contributions of Heisenberg, Schrödinger,
Born, Pauli, Dirac, and Einstein himself were of course crucial, particularly as concerns the physics and mathematics of quantum theory. At
the philosophical level, however, Bohr's role was unrivaled and remains
unsurpassed.

Bohr appeared on the scene of quantum physics twice in major roles, first in 1913, with his theory of the atom — a brilliant application of Planck's quantum principle — and then around 1927, with the principle of complementarity grounding his interpretation of quantum mechanics.[13] Bohr developed the principle and then the framework of complementarity from what he defined as *complementary* features, that is, features which are mutually exclusive (and thus not simultaneously applicable at any given point) but equally necessary for a comprehensive description of quantum phenomena. As an overall framework, complementarity entails both an irreducible division and an irreducible interaction between, on the one hand, quantum (micro) objects, described by means of *quantum-mechanical* formalism, and, on the other hand, a measuring (macro) apparatus, described by means of the formalism of *classical* physics. It is through this interaction that complementary descriptions themselves emerge, entailed by mutually exclusive experimental arrangements; and, according to Bohr, if the mathematical formalism of quantum mechanics *represents* anything at all, it is this *interaction* rather than quantum objects themselves (or what are inferred as such in the process). These circumstances already indicate difficulties in developing a realist interpretation of quantum phenomena, since at any given point at least two, mutually exclusive "pictures" are always possible and there appears to be no underlying (classically) complete configuration that such pictures would partially represent. Such pictures can even be multiple, that is, entailing more than two complementary pictures, for example, if one constructs different complementary representations along different axes in space. The overall situation is clearly at odds with classical realist conceptualizations.

Two forms of complementarity are of particular significance in Bohr's framework. The first is the complementarity of *wave* and *particle*, in which the duality of quantum phenomena is reflected and the continuous and discontinuous representations of quantum processes are related. These two phenomena are unequivocally dissociated in classical physics, which distinguishes between discontinuous, particle-type phenomena, such as electrons; and continuous, wave-type phenomena, such as light or other forms of radiation. In quantum physics, however, such unequivocal identification becomes impossible. Light, classically a wave phenomenon, acquires a double nature, or, more precisely, it requires two complementary modes of representation via mutually exclusive experimental arrangements: at times, depending on what experimental arrangements we make, light must be represented as particles — photons — and at other times as waves, but never as both simultaneously. While there have been classical corpuscular theories of light, such as Newton's, prior to quantum mechanics, light and matter alike were always subject to one form of representation or the other, never to a combination of both. In contrast, quantum physics must employ both types of representation, which reflect the mutually exclusive experimental arrangements involved. This understanding does not imply either that experiments can fully *create* or control what is observed or, conversely, that one needs to postulate something that exists in itself, independently of observation. Bohr rejected both views as untenable under quantum conditions. One can only argue that technology, including the "technology" of our perception, creates *specific* complementary pictures, depending on the "arrangements" we make. Our perception appears to be classical—that is, it appears to generate pictures conforming to classical physics—which fact may be ultimately responsible for complementarity. (I shall consider this situation in more detail later.)

The second form of complementarity is that of coordination and causality. It dislocates the causal dynamics that define the behavior of classical systems and allow one to know with certainty the state of a given system at any point. "The very nature of the quantum theory," explained Bohr, "forces us to regard the space-time co-ordination and the claim of causality, the union of which characterizes the classical theories, as complementary but exclusive features of the description, symbolizing the idealization of observation and definition respectively."[14] The complementarity of coordination and causality is directly connected to the complementarities of position and momentum and of time and energy, which preclude one from simultaneously measuring both variables within each pair, or, according to Bohr, even from meaningfully considering them as simultaneously applicable or definable at any point. This impossibility arises because again one needs mutually exclusive experimental arrangements for the strict measurement, applicability, and definition of each such variable.[15] The mathematical counterpart of the complementarity of coordination and causality becomes Heisenberg's uncertainty relations, which express the limits on the possibility of (this is crucial) *simultaneous* exact measurement of such complementary (or, as they are also sometimes called, conjugate) physical variables.

Complementarity and the conflict with the classical description that it entails were seen by Bohr as inevitable consequences of Planck's quantum postulate:

> The quantum theory is characterized by the acknowledgment of a
> fundamental limitation in the classical physical ideas when applied

to atomic [i.e., quantum] phenomena. The situation thus created is of a peculiar nature, since our interpretation of the experimental material rests essentially upon the classical concepts. Notwithstanding the difficulties which, hence, are involved in the formulation of the quantum theory, it seems . . . that its essence may be expressed in the so-called quantum postulate, which attributes to any atomic process an essential discontinuity, or rather individuality, completely foreign to the classical theories and symbolized by Planck's quantum of action.[16]

The concluding formulation in this passage is crucial. Reinforced by similar propositions throughout Bohr's writing, it tells us that, along with continuity, quantum discontinuity (or even complementarity) has only a symbolic character. It is a symbolization and an idealization rather than a (classically conceived) representation of any (classically conceived) physical reality, independent properties of physical objects, and so forth. Both terms ("symbolization" and "idealization") and the concepts they designate, like such terms as "abstraction" (also used by Bohr) and the more current "metaphorization," have a long history, reflecting the roles they have played, sometimes simultaneously, in both the solidification and the dislocation of classical thinking. In Bohr's writing, these terms operate mainly in the latter capacity. This dislocating force emerges more clearly from his argument than from any explicit definitions; and, as in other postclassical theories, these terms do not always permit strict definition. At an initial level of approach, one might see "idealization" and related terms used by Bohr as referring to operational constructs, which are interactively mental and linguistic or technological—such as traces left in cloud chambers or on photographic plates, or images produced on computer terminal screens, virtual reality devices, and so forth. Such constructs enable (but never fully determine) our theories and communication in either more or less formal or more or less informal modes. The formal and specifically mathematical aspects of such concepts are, of course, irreducible in physics. We can, however, be no more absolutely formal than absolutely informal, which latter fact may be less obvious but is no less crucial. Bohr stressed our dependence on "everyday" concepts and language, on pictures, metaphors, visualization, and so forth, along with physical and mathematical concepts. By using such formal and informal constructs, we can relate—by definition, obliquely—to the radical, but never absolute, alterity encountered in quantum physics.

All the concepts through which we approach quantum physics are idealizations in the sense just delineated, however much they are enabled by the experimental technology at our disposal; and experimental technology enables conceptual idealizations in all physics, classical and quantum alike. There may be no nontechnological idealization anywhere, as postclassical theories suggest, and these theories transform the very concept of technology in turn. Bohr's analysis proceeded by exposing and using the fundamental connections between the mathematical formalism of quantum mechanics and the experimental arrangements involved in obtaining its data. He argued that the only, in his terms, rational (i.e., comprehensive and consistent) interpretation of quantum data fundamentally, irreducibly connects *quantum* formalism, interestingly, to the *classical* description of the measuring instruments through which we obtain the data. These data allow us to infer the existence of quantum objects. The character of these data, however, is such that no realist interpretation appears to be possible. Instead, what I call here radical yet reciprocal alterity appears to be necessary, along with the codependence that arises between what is observed and the agencies of observation:

> Our usual description of physical phenomena is based entirely on the idea that the phenomena concerned may be observed without disturbing them appreciably. . . . Now, the quantum postulate implies that any observation of atomic phenomena will involve an interaction with the agency of observation not to be neglected. Accordingly, *an independent reality in the ordinary physical sense can neither be ascribed to the phenomena nor to the agencies of observation.* After all, the concept of observation is in so far arbitrary as it depends upon which objects are included in the system to be observed. Ultimately, every observation can, of course, be reduced to our sense perceptions. The circumstance, however, that in interpreting observations use has always to be made of theoretical notions entails that for every particular case it is a question of convenience at which point the concept of observation involving the quantum postulate with its inherent "irrationality" is brought in.
>
> This situation has far-reaching consequences. On one hand, the definition of the state of a physical system, as ordinarily understood, claims the elimination of all external disturbances. But in that case, according to the quantum postulate, any observation will be impossible, and, above all, the concepts of space and time lose their immediate sense. On the other hand, if in order to make observation possible we permit certain interactions with suitable

agencies of measurement, not belonging to the system, *an un-ambiguous definition of the state of the system is naturally no longer possible, and there can be no question of causality in the ordinary sense of the word.*[17]

The propositions I stress here — "*an independent reality in the ordinary physical sense can neither be ascribed to the phenomena nor to the agencies of observation,*" and, as a result, "*an unambiguous definition of the state of the system is naturally no longer possible, and there can be no question of causality in the ordinary sense of the word*" — define central postclassical features of Bohr's framework, linked by him to the interaction between quantum objects and classical measuring instruments. Similarly, in the case of wave-particle complementarity:

> The two views of the nature of light are . . . to be considered as different attempts at an interpretation of experimental evidence in which the limitation of the classical concepts is expressed in complementary ways. The problem of the nature of the constituents of matter presents us with an analogous situation. . . . Just as in the case of light, we have consequently in the question of the nature of matter, so far as we adhere to classical concepts, to face an inevitable dilemma which has to be regarded as the very expression of experimental evidence. In fact, here again we are not dealing with contradictory but with complementary pictures of the phenomena, which only together offer a natural generalization of the classical mode of description. . . . [It] must be kept in mind that . . . *radiation in free space as well as isolated material particles are abstractions, their properties on the quantum theory being definable and observable only through their interaction with other systems.* Nevertheless, these abstractions are . . . indispensable for a description of experience in connection with our ordinary space-time view.[18]

Thus, wave-particle complementarity, too, is defined by relating it to specific experimental arrangements in which specific observations are made and in relation to which specific questions are asked. This complementarity can be related to anti-causality through Max Born's interpretation of Schrödinger's wave-function as mapping the probability distributions that define the predictions of quantum mechanics. Born's interpretation suspends the physical significance of quantum waves. In Bohr's still more radical view, an independent physical reality cannot be ascribed to either particles or waves, making the application of the concept of physical reality to quantum physics deeply problematic.

Bohr developed complementarity into a comprehensive framework that encompassed both quantum physics and *meta*-physics, which term we may distinguish here from *metaphysics*. Bohr's meta-physics is *antimetaphysics*, at least in relation to metaphysics as developed by Western philosophy from Plato and Aristotle on. Meta-physics, by contrast, would refer to extra-physical considerations, which may proceed by means of antimetaphysical theories, such as Bohr's or, for that matter, Nietzsche's and Derrida's. Heisenberg reported the following remarks by Bohr on Philipp Frank's lecture in which Frank "used the term 'metaphysics' simply as a swearword or, at best, as a euphemism for unscientific thought":

> I began by pointing out that I could see no reason why the prefix "meta" should be reserved for logic and mathematics — Frank had spoken of metalogic and metamathematics — and why it was anathema in physics. The prefix, after all, merely suggests that we are asking further questions, i.e., questions bearing on the fundamental concepts of a particular discipline, and why ever should we not be able to ask such questions in physics?[19]

These questions, however, can be asked, as they were by Bohr, in terms of postclassical theories. Such theories may be seen as analyses of the nonclassical *efficacity* (which term I shall oppose here to causality) of processes that classical theories understand by such terms as causality, or meaning, presence, and truth, or full consciousness and self-consciousness, or the possibility of unconditionally grounding or centering interpretation or theory, or the possibility of a fully coherent unity of knowledge, or the possibility of projects based on these assumptions, and so forth. While classical theories may differ, sometimes significantly, from each other, they are complicit within what Derrida designates as the metaphysics of presence — a powerful and systematic unity that has evidently defined the history of Western philosophy, but is operative well beyond its limits. Derrida also uses such terms as "logocentrism" and, following Heidegger, "ontotheology" (the latter connoting the complicity of philosophical and theological thinking). He associates Nietzsche's thinking, especially his concept of *play*, with "the shattering [*ébranlement*] of ontotheology and the metaphysics of presence."[20] Bohr's complementarity achieved comparable results in relation to the major philosophical assumptions defining classical thinking in physics and the philosophy of physics.

In his account of the genesis of uncertainty relations, Heisenberg describes his thinking at the time as follows:

There was not a real path of the electron in the cloud chamber. There was a sequence of water droplets. Each droplet determined inaccurately the position of the electron, and the velocity could be deduced inaccurately from the sequence of droplets. Such a situation could actually be represented in the mathematical scheme; the calculation gave a lower limit for the product of the inaccuracies of position and momentum [i.e., uncertainty relations].[21]

As will be seen, Heisenberg's formulation is not sufficiently precise, and his views were criticized by Bohr, whose own understanding of this situation led him to complementarity. It is clear from Heisenberg's description, however, that in quantum physics one deals only with traces and, finally, with traces of traces, which suggests that our interpretation of experimental data may entail what Derrida describes as "the strange structure of the supplement," where "by delayed reaction, a possibility produces that to which it is said to be added on."[22] According to Bohr, our inferences concerning the *efficacity* (as opposed to the *causality*) of quantum data no longer allow us to infer a classical or classical-like configuration that would be *represented*—directly or indirectly, completely or incompletely (for example, statistically)—by these data. Such data instead require complementarity. The quantum mechanical situation does have its own specificity. Indeed, while possessing features, such as supplementarity, which it shares with other postclassical frameworks, in some of its key aspects quantum physics is quite unique—without precedent anywhere else. One of Bohr's great achievements was to understand and theorize this specificity, which he does in terms of *complementarity*. Nevertheless, it is worth spelling out the key general features of *supplementarity*.

Supplementarity does not imply that a trace, one left by an electron or any supplementary mark, for example, precedes rather than follows that of which it is a trace or mark. Supplementarity, that is, does not simply reverse a classical order of causality or signification. A trace or mark is neither simply added to the efficacity of this trace, nor does it simply produce or even condition this efficacity. Instead, to the extent that one can speak in the classical terms here, both a trace and that of which it is a trace may be said to be affected and effected by each other within an interpretive process. What marks and what is marked supplement each other, making Derrida's supplementarity correlative to Nietzsche's critique of classical causality.[23] Derrida himself approaches such reciprocal efficacities through the concept of supplementarity and related theoretical structures, such as *différance, trace, writing,* and so forth. These efficacies cannot be conceived of in terms of some original presence, even if the latter is seen as displaced by a representation. Both in quantum mechanics and in Derrida's framework, "nothing, neither among the elements nor within the system, is anywhere ever simply present or absent. There are only, everywhere, differences and traces of traces."[24] One cannot, however, think of this dynamic in terms of efficacies that presuppose an original presence of which a trace would be simply a trace or a trace of a trace, however remote such a presence might be. One must instead think simultaneously in terms of trace-like movement and of supplementarity. Both together displace classical conceptions of absolutely original structures, whether empirical or transcendental, from which something could be thought to differ or in relation to which something could be seen as delayed. Any claim concerning the nature (e.g., a causal or, conversely, acausal nature) of this efficacy can be supplementarily derived only from the latter's effects or traces via mathematical, technological, conceptual, metaphorical, or institutional processes. Supplementarity does not imply that there is no material efficacy to quantum traces, but rather that such materiality cannot be conceived of classically — which, of course, entails a revision of the classical concept of matter.

Given its impact on modern intellectual history, quantum mechanics may well have figured among Derrida's sources. In his writing, the analysis of supplementarity proceeds mainly via Nietzsche and Freud and the deconstruction of classical conceptions of causality and signification, most specifically as they appear in texts by Rousseau, Saussure, and Husserl, but also as they feature implicitly throughout the history of Western philosophy.[25] As a dislocation of classical causality, however, supplementarity is a defining aspect of much postclassical thinking. Thus, the (ex post facto) transformational dynamics of Darwin's theory of evolution and subsequent theories in biology, most recently Gerald Edelman's neural Darwinism, can be shown to be fundamentally supplementary.

Bohr's complementarity belongs to and extends the same multiply parallel and multiply interactive network of postclassical thinking, whose history must itself be understood in supplementary — and complementary — terms.[26] According to Bohr, "the finite [quantum] and uncontrollable [indeterminate] interaction" between quantum objects and measuring instruments" makes it no longer possible to apply classical theories and pictures, except by means of complementarity.[27] In contrast to classical physics, the intervening process of this interaction can no longer be disregarded, which impossibility entails complementarity.

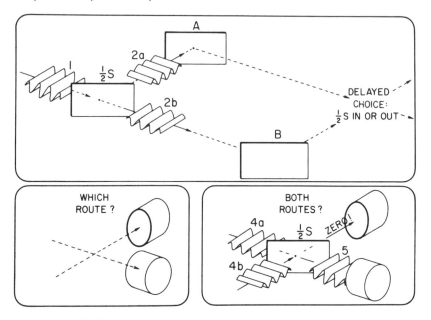

Figure 1. The beam splitter experiment. From John Archibald Wheeler and Wojciech H. Zurek, eds., *Quantum Theory and Measurement* (Princeton, 1983), 183. Copyright 1983 by Princeton University Press. Reproduced by permission of Princeton University Press.

A variety of actual and thought experiments have been designed to illustrate the situation, and most books on the subject contain descriptions of them. For example, John Archibald Wheeler provides the following account and graphic depiction (Figure 1) in "Law without Law":

> An electromagnetic wave comes in at 1 and encounters the half-silvered mirror marked "½S" which splits it into two beams 2a and 2b, of equal intensity which are reflected by mirrors A and B to a crossing point at the right. Counters (lower left) located past the point of crossing tell by which route an arriving photon has come. In the alternative arrangement at the lower right, a half-silvered mirror is inserted at the point of crossing. On one side it brings beams 4a and 4b into destructive interference, so that the counter located on that side never registers anything. On the other side the beams are brought into constructive interference to reconstitute a beam, 5, of the original strength, 1. Every photon that enters at 1 is registered in that second counter in the idealized case of perfect mirrors and 100 per cent photodetector efficiency. In the one

arrangement (lower left) one finds out by *which* route the photon came [in a particlelike manner]. In the other arrangement (lower right) one has evidence that the arriving photon came by both routes [in a wavelike manner].[28]

There is no situation in which a photon can simultaneously behave both as a particle and as a wave, nor is there a single "picture" that synthesizes both patterns. This circumstance, according to Bohr, demanded a radical revision of our understanding of the problem of physical reality. Bohr's argument was directed in particular against Einstein, whose philosophy of physics can be summarized in the following four principles:

(1) the *causality* of and absence of randomness in all physical interactions — at bottom, if not in all overt manifestations, which may be statistical;

(2) the *continuity* of mathematical representations involved, beginning with the space-time continuum;

(3) the *completeness* of theory, defined in classical terms, via Einstein's concept of a physical reality;

(4) the concept of physical *reality* as existing independently of observation and to which physical theory must conform.

All of these features are fundamentally threatened by quantum mechanics. As Wheeler notes,

The dependence of what is observed upon the choice of experimental arrangement made Einstein unhappy. It conflicts with the view that the universe exists "out there" independent of all acts of observation. In contrast Bohr stressed that we confront here an inescapable new feature of nature, to be welcomed because of the understanding it gives us. In struggling to make clear to Einstein the central point as he saw it, Bohr found himself forced to introduce the word "phenomenon." In today's words Bohr's point — and the central point of quantum theory — can be put into a single, simple sentence. "No elementary phenomenon is a phenomenon until it is a registered (observed) phenomenon." It is wrong to speak of the "route" of the photon in the experiment of the beam splitter. It is wrong to attribute a tangibility to the photon in all its travel from the point of entry to its last instant of flight. A phenomenon is not yet a phenomenon until it has been brought to a close by an irreversible act of amplification such as the blackening of a grain of silver bromide emulsion or the triggering of a photo-detector. . . . What answer we get depends on the question we put, the experiment we arrange, the registering device we choose.[29]

The situation manifests itself even more dramatically in the so-called delayed-choice version of the experiment described above. According to Wheeler,

> In the new "delayed-choice" version of the experiment one decides whether to put in the half-silvered mirror or take it out at the very last minute [i.e., the very last picosecond]. Thus one decides whether the photon "shall have come by one route, or by both routes" after it has "*already done* its travel." . . . In this sense, we have a strange inversion of the normal order of time. We, now, by moving the mirror in or out have an unavoidable effect on what we have a right to say about the *already* past history of that photon."[30]

One cannot, however, speak of dynamical properties and trajectories of particles prior to an act of observation, upon the results of which such classical concepts can be (supplementarily) superimposed in one way or another, classically or postclassically. As Wheeler writes:

> In the delayed-choice version of the split-beam experiment, for example, we have no right to say what the photon is doing in all its long course from point of entry to point of detection. Until the act of detection the phenomenon-to-be is not yet a phenomenon. We could have intervened at some point along the way with a different measuring device; but then regardless whether it is the new registering device or the previous one that happens to be triggered we have a new phenomenon. We have come no closer than before to penetrating to the untouchable interior of the phenomenon. For a process of creation that can and does operate anywhere, that reveals itself and yet hides itself, what could one have dreamed up out of pure imagination more magic — and more fitting — than this?[31]

Wheeler's choice of the expression "the phenomenon-to-be" is very apt. "The phenomenon-to-be" is what will be registered as a trace or effect at a certain point, but the efficacity of which can never — not in the past, the present, or the future — be seen as *present*, whether as representable or as unrepresentable, existing in itself or by itself, independent of and inaccessible to any interpretation. This efficacity is, in these respects, analogous to the Derridean supplementary efficacity, which "makes us concerned not with horizons of modified — past or future — present, but with a 'past' that has never been present, and which never will be, whose future to come [*l''a-venir'*] will never be a *production* or a reproduction in the form of presence."[32] All of our "photographs" are taken "too late" to allow us to describe or even to speak of a reality in the sense of an autonomous presence behind quantum phenomena. From the classical point of view, these phenomena are incomplete and contradictory, but in any case the degree to which one could speak in classical terms at this point would be very limited. In particular, it is not a question of somehow modifying a past that has already existed independently of a later interpretation. Were such the case, Wheeler's claims would lead to logical problems which Bohr's interpretation avoided.

The delayed-choice experiment can even be enacted on a literally cosmic scale and is a very dramatic illustration of the ineluctable strangeness of the quantum world.[33] Fundamentally, however, it is no different from other quantum experiments, as Bohr realized.[34] It merely amplifies the fact that in quantum physics one deals with an irreducible (or, as Derrida calls it, an "absolute") past — a past that is incompatible with the category of present, whether as now-present, past-present, or future-present.[35] The "finite and uncontrollable" interaction with measuring devices irreducibly and uncontrollably displaces and defers the efficacity of the quantum phenomena — whether particles or waves, or such dynamical variables as position and momentum — registered by those measuring devices. It places this efficacity into an irreducible past and makes of it a certain radical — but again not absolute — alterity. This irreducible past suspends both the possibility of anything's existing by and in itself, independently of interpretation, and the possibility of classical representation, full or even partial, insofar as the latter is seen as capturing, directly or indirectly, some elements of an independently existing reality. It follows that no mathematical, conceptual, or metaphorical model — continuous, discontinuous, or complementary — can be assigned the status of physical reality in any classical sense: there is no conglomerate of properties that can be postulated as even partially representable or displaced by means of such models or any other models. Nor can one speak of reality as existing by itself, either alongside its partial or displaced representations or as absolutely inaccessible to any representation. Hence Bohr argued, against Einstein, that one cannot assign independent properties to quantum objects (or what we infer to be such from our observations). As he said, "It is difficult for me to associate any meaning with the question of what is behind the phenomenon."[36] Bohr rejected the notion of either "creation" or "disturbance" through observation:

> I warned especially against phrases, often found in the physical literature, such as "disturbing of phenomena by observation" or "creating physical attributes to atomic objects by measurement." Such phrases, which may serve to remind us of the apparent paradoxes in quantum theory, are at the same time apt to cause confusion, since words like "phenomena" and "observation," just as "attributes" and "measurements," are used [here] in a way hardly compatible with common language and practical definition.[37]

A suspension of these and other *classical* possibilities does not, however, exclude all *theoretical* possibilities. Quantum mechanics and post-classical theories elsewhere are able to operate productively with effects (classical or otherwise) of efficacities that are not themselves accessible to any classical conceptualization or, conceivably, to any conceptualization available to us, again including those in which such efficacities appear to be absolutely inaccessible. Nor does it follow that the efficacity theorized here via radical alterity does not exist or does not produce *material* traces as its effects. What follows is only that this efficacity cannot be defined by means of any mode or concept of existence available within our interpretive and theoretical enclosure or enclosures, classical or postclassical. The materiality of this efficacity does, of course, problematize all classical *phenomenality*—Kantian, Hegelian, Husserlian, or other. It makes equally problematic, however, all concepts of reality hitherto conceived, including those posited or implied by metaphysical materialism or objectivism. For the matrix entailed by this efficacity cannot be seen as "objective" either, given the history of that term and concept. The choice or necessity of any given term is always strategic and historical. The term "objective" is problematic, even though Bohr himself used it sometimes. His emphasis, however, was on *"conditions for objective description."*[38] Rather than implying objectivity or reality in the classical sense, he refers to the conditions for the possibility of unambiguous communication of experimental results; the possibility of making experimental results sufficiently independent of a particular observer; and the impossibility of any observer's influencing the results under arranged conditions — all of which can be retained in a postclassical framework — Bohr spoke of "a radical revision of our attitude toward the problem of physical reality," and "problem" may well be the key word here.[39]

This radical revision entails the indeterminacy, as opposed to the un-equivocal definability, of the relationships between the "exterior" and the "interior" of any quantum event (such terms being preferable to "subject" and "object" under these conditions). The very division, or so-called cut, between that which observes and that which is observed is arbitrary in quantum mechanics. The cut may separate either the "original" quantum object from the "original" classical instrument or the object combining this object and this instrument from some further instrument, and so on. As Bohr argued, quantum indeterminacy affects the (quantum) interaction between measured objects and measuring instruments regardless of where the cut is made because both must be treated at the microlevel as quantum systems, even though at the macro-level instruments must be treated (and all data observed) classically.[40] Indeterminacy itself remains irreducible under all conditions.[41]

Complementarity is a materialist theory in the sense that it assigns a material character to the alterity-efficacity it considers, which could be seen as defining matter itself, except that matter can no longer be said to exist "by itself" or "in itself." Derrida speaks (via Nietzsche and Bataille) of matter as radical alterity: "If, and in the extent to which, *matter . . .* designates . . . radical alterity (I will specify: in relation to philosophical oppositions), then what I write can be considered 'materialist.' "[42] Else-where, he defines radical alterity even more strongly: it "concerns every possible mode of presence."[43] It is clear that this alterity cannot be seen as absolute, since it would then be defined by a metaphysical opposition of absolute identity and absolute difference (or other oppositions of this type). Nor can this alterity be governed by any single structure, name-able or unnameable, or by a containable cluster of structures. While this alterity-efficacity cannot be configured in classical terms — or, *at the limit,* in any terms — it produces all the *effects* described via such terms, making our interpretation and theory, including our approaches to this alterity, depend on these effects. This dependence cannot, how-ever, be subject to any classical (classically causal or classically statistical) understanding.

One could attempt to define "reality" by way of, or precisely as, radical alterity. The latter, however, does not appear to correspond meaningfully to any conception of reality hitherto conceived, which explains in part why the term "reality" and the concepts that it designates have so far inhibited rather than enhanced postclassical possibilities for the inter-pretation of quantum physics and of Bohr's complementarity. "Alterity" appears to be, conceptually and rhetorically, a more effective term than "reality" would be, but alterity (or, for that matter, "efficacity") is not meant to be the governing or sole term for the efficacity or efficacities at issue here. Instead, the conception of alterity here is meant to minimize metaphysical claims upon this efficacity. In fact, this conception makes no claims conforming to the metaphysics of presence as hitherto constituted, including its constitution as scientific empiricism or objectivism. By retaining (a certain) materiality, however, postclassical theories allow one to deconstruct the opposite — that is, anti-materialist — forms of metaphysics, such as the idealist, phenomenologist, or positivist forms.

Of course, any given term, including "matter" or "alterity," may pose problems of reinstating the metaphysics of presence. To avoid these problems, one must carefully follow the postclassical logic exemplified in Bohr's work or elsewhere. This logic appears to tell us that the alterity-efficacity at issue here may not conform to any concept (even where it is, as in Derrida, "neither a word nor a concept") or to any conceptual or interpretive formation that is or ever will be available to us within the conceptual and interpretive (including the material and technological) enclosures where we must function. The "alterity" at stake may not conform to anything that can be approached through any terms, including those of being/becoming, beginning/ending, difference/exteriority (or, of course, identity/interiority), unity/multiplicity, finitude/infinity, and matter/spirit. By the same token, these considerations do not suggest, nihilistically, that the "process(es)" and "relation(s)" at issue relate to nothingness. Here, they must not — and cannot — relate to what may be no more "nothing" than "something," or any other "thing," to what is no more "it" than unidentifiable by this or any pronoun, to what *is* no more than *is not*, but that nevertheless is what produces the effects of nothingness or "somethingness" (or "thingness"), of being and nonbeing, and all the other effects just indicated and all the terms used here. Although such a "relation" is classically impossible — and, conceivably, impossible or *inconceivable* in any way — at issue is a "relation" or "non-relation" of neither absolute (or potentially any) alterity, difference, exteriority, or disconnection; nor of absolute (or potentially any) connectibility, or anything within any form of the conceptual and interpretive enclosure within which we must function. The latter formulation, of course, also suggests a form of relation, and, as such, it may be equally inapplicable. It follows, in fact, that all propositions offered at the moment — *now, in this text* — may be equally inapplicable. Furthermore, this "relation" "relates" to (with these and all subsequent terms in quotation marks), or "disconnects" from, a "process" that may be neither being nor becoming, neither one nor many, neither finite nor infinite, and that neither has origins or ends or middles nor conforms to any other conceivable characterizations — whether in terms of a classical or post-classical theoretical economy, or by means of any model — including, again, any characterizations now offered here. All of our "relations" to this alterity are subject to the same economy. We may not be able either to connect or to disconnect ourselves with or from this alterity through any interpretation, theory, or technology that is or ever will be available to us (if, like other terms used here, or again like any terms, these terms are applicable). The very concept of the *impossibility* of relating this alterity to any interpretation may be inapplicable, so the last proposition — or this very proposition or, again, any other negating or qualifying proposition of that type — may itself need to be negated (or double-negated) without returning to what it negates.

Is it possible to conceive of something that neither connects nor disconnects or is neither connected nor disconnected, something that is indeed neither conceivable nor inconceivable, never fully, or even to any degree, inside or outside any enclosure, given that all such questions are themselves the products of enclosures, beyond some of which one must move and within others of which one must remain in order to approach this radically, but never absolutely, unrepresentable or unidealizable process? It may or may not be possible ever to do so, as the very concepts of conceptualization, representation, and idealization — or of possibility — may need to be further scrutinized, transformed, or abandoned. This alterity, however, appears to be unavoidable in the situation that one encounters in quantum physics.

As a correlative of complementarity, Heisenberg's uncertainty relations express mathematically the limits on the possibility of the *simultaneous* exact measurement — and, according to Bohr, the ascription and definition — of complementary or conjugate variables, such as position and momentum or time and energy. The term "conjugate variables" originates in classical mechanics, where such variables can be connected to each other by means of so-called Hamiltonian formalism, which represents classical mechanics (originally defined by Newton's laws of motion) in terms of differential equations. These connections between mathematics and physics, of course, played a major role in Newton's invention of differential calculus and had a great impact on the subsequent history of physics, including Hamilton's thinking. In a way, what Hamilton did for classical mechanics was what Heisenberg or Schrödinger did later for quantum mechanics by using a very different mathematics or, one might say, by translating Hamiltonian formalism into quantum-mechanical mathematical formalism, with a rather extraordinary postclassical outcome. Hamilton developed his formalism by exploring connections between mathematical descriptions of dynamical and optical processes, and the rise of quantum mechanics gave new and profound significance to his discovery.[44] The mathematical formalism of quantum mechanics, from which uncertainty relations developed, may be seen as a kind of translation of this classical formalism into the language of a different mathematical theory — that of the so-called Hilbert spaces. By the same token, this latter mathematical theory reflects a very different physics, one that dislocates the possibility of realist interpretations entailed, or at least permitted, by classical physics. Bohr connected both formalisms — the classical (Hamiltonian) and quantum-mechanical (defined in terms of Hilbert spaces) — by interpreting quantum-mechanical formalism as "representing" or, again, idealizing the interaction between measuring instruments, described in terms of *classical* physics, and *quantum* objects (or, again, what are inferred to be such objects from such interactions). The role of this interaction is irreducible, and a full determination of one conjugate or complementary variable always requires an experimental arrangement which makes it impossible to fully determine the other. According to Bohr, uncertainty relations reflect this mutually exclusive character of the two arrangements, which are themselves described in terms of classical physics — but only *at the macrolevel*. At the *microlevel* measuring instruments must, of course, themselves be seen as quantum or at least as capable of quantum ("finite and uncontrollable") interactions with quantum objects. This double describability of measuring instruments (classical at the macrolevel, quantum at the microlevel), central to Bohr's interpretation, was what allowed him to connect the mathematical formalism of quantum mechanics and quantum measurement.

In the mathematical formalism of quantum mechanics, variables (or observables, as they are also called) are represented by what are called operators in a Hilbert space — P (momentum), Q (coordinate), and so forth — to which standard mathematical operations, such as addition and multiplication, can be assigned. Such spaces are complex mathematical objects, with an infinite number of dimensions; operators represent transformations of the elements of such spaces. The actual mathematics of Hilbert spaces need not be considered here. The key point is that, while this type of formalism enables one to account for the experimental data, it does not represent any *actual* or presumptive (independent) physical properties of quantum objects, which properties, as we have seen, cannot even be spoken of in quantum mechanics. In classical physics, by contrast, variables are represented as standard mathematical functions of time and of each other. For example, the energy of a system would depend on the momenta and coordinates of its components. Classical conjugate variables like position and momentum are such functions, again connected by means of Hamilton's equations. The latter represent mathematically the fact that the change of one conjugate variable — say, position — in time corresponds to a change in the energy of the system with respect to the other conjugate variable — say, momentum. The respective evolutions of both variables in time become related (hence "conjugate"). This is also true in quantum mechanics, with the difference that here *complementary* conjugate variables acquire the uncertainty that makes these variables complementary, for this uncertainty affects — that is, precludes — the possibility of their joint unambiguous determination and thereby, in contrast to classical physics, fundamentally disallows a realist interpretation of quantum physics. What forestalls such an interpretation is a peculiar feature of this mathematical formalism: that is, the multiplication of such operators in general does not commute, specifically when conjugate variables are involved — PQ does not equal QP.[45] This noncommutativity represents the fact that corresponding variables, such as momentum and position, cannot be fully determined and defined simultaneously. The difference between the two products, PQ and QP, allows one to establish quantitative limits, in this respect represented by Heisenberg's uncertainty relations. This noncommutability translates, as it were, the relation between conjugate variables of classical physics, but in such a way that no increased degree of definition in one such variable can be achieved without a loss to the same degree of definition in the other. The "state" of a quantum system may be described by another mathematical object (an element rather than an operator) in a Hilbert space associated with the system: a so-called state vector. The latter, however, again reflecting quantum indeterminacy, is defined so as to prevent one from understanding the state of the system on the classical model, in which the joint determination of variables and of the overall dynamics of the system is possible (i.e., is not subject to indeterminacy).

It is important that quantum-mechanical formalism limits only the *simultaneous joint* measurement, applicability, and definability of conjugate variables within the *same* experimental arrangement. Either variable by itself can always be measured with the full precision available to us and may be well defined at any given point. The latter fact can make it tempting to think, as did Einstein, that something is wrong with quantum-mechanical description, that it is somehow incomplete and that one can develop a more complete classical-like account of the data at issue. According to Bohr, however, these facts, while reflecting a peculiar, even unique, strangeness of the quantum world, point to the inapplicability of classical physical and philosophical ideals to quantum physics.

Of course, there is always a degree of uncertainty or indeterminacy in measuring a single variable—say, coordinate—by itself as well. It appears, however, that we can measure any such variable by itself with any desired degree of accuracy. That is, this accuracy is constrained by the technical capabilities of our measuring instruments at the moment rather than by any fundamental limitation. It is only when one needs to determine both conjugate variables simultaneously that fundamental, uncircumventable limits arise, regardless of the capabilities of our instruments, and it is these irreducible limits that are at stake in uncertainty relations.

The main reason for this limitation is, again, that respective full measurements and unambiguous definitions of each variable would require mutually exclusive experimental arrangements. When one considers a trace left by a particle (say, the "sequence of water droplets" of Heisenberg's description cited earlier [46]), one does, in a certain sense—and, in classical terms, "inaccurately"—determine or, according to Bohr, even assign both variables. Thus Heisenberg could say there that "each droplet determined inaccurately the position of the electron, and the velocity [or momentum] could be deduced inaccurately from the sequence of droplets. . . . A lower limit for the product of the inaccuracies of position and momentum" was given by uncertainty relations. Heisenberg's formulation was not altogether precise, however, and Bohr criticized his initial interpretation of uncertainty relations (via "Heisenberg's microscope") for this lack of conceptual precision. It suggested that there might be a classical-like trajectory which our measurements distort and which, as Einstein intimated, quantum formalism incompletely represents. Since this is not so, however, Heisenberg was right to say that "there was not a real path of the electron in the cloud chamber," which point became crucial for Bohr. As he showed, the experimental arrangements necessary for controlling the indeterminacy in measuring one conjugate variable with unlimited precision—and such arrangements are or at least appear to be always possible, at least in principle—would entail renouncing the possibility of controlling the indeterminacy in measuring the other variable. Once we control the indeterminacy in measuring one complementary variable, as we can always do (at least to any limits hitherto available), we unavoidably lose our physical means of controlling the indeterminacy in measuring the other, and there is no way to circumvent this mutual exclusivity.[47] Hence quantum mechanics must be seen, according to Bohr, as a *complete* description, as complete as it can be, given quantum conditions.[48] In classical terms, the picture is incomplete, and uncertainty relations may be said to measure this incompleteness. According to Bohr, however, they signal the ultimate inapplicability to quantum mechanics of a classical-like interpretation, whether strictly causal or classically statistical (implying an underlying nonstatistical configuration incompletely represented by a statistical account). As he observed,

> These circumstances find quantitative expression in Heisenberg's indeterminacy relations which specify the reciprocal latitude for the fixation, in quantum mechanics, of kinematical [position] and dynamical [momentum] variables required for the definition of the state of a system in classical mechanics. In fact, the limited commutability of the symbols by which such variables are represented in the quantal formalism corresponds to the mutual exclusion of the experimental arrangements required for their unambiguous definition. In this context, *we are of course not concerned with a restriction as to the accuracy of measurement,* but with a limitation of the well-defined application of space-time concepts and dynamical conservation laws, entailed by the necessary distinction between [classical] measuring instruments and atomic [quantum] objects.[49]

The concluding, rather striking, sentence here defines Bohr's understanding of observation and measurement in quantum physics. *Classical* instruments show us only *classical* "pictures" — a classical wave-picture or a classical particle-picture (but never both together); a classical picture of the momentum of a particle or a classical picture of its position (but never both together); and so forth — some of which can be correlated. This makes sense. *Both* our instruments *and* the pictures they produce are classical, and we can have no other instruments and can "see" no other pictures. We infer quantum objects from the "sum" of these pictures and by relating and correlating some of them. These pictures, however, never quite "sum up," certainly not in the way that they do in classical physics. There is no underlying complete classical picture from which these complementary classical "parts" can be derived, which impossibility is the essence of quantum mechanics and complementarity. One cannot even say that the whole is more (or less) than the sum of its parts. There *is no* "whole" that is the sum of these "parts," which makes the applicability of such terms (and, again, possibly any terms available to us) problematic.

In contrast to classical physics, the interaction between quantum objects and measuring instruments in quantum mechanics both entails an indivisible or interactive whole and requires a strict discrimination between them, insofar as the description of their physics is concerned. This joint structure of necessary interconnection and necessary difference in description may appear paradoxical, but it is a logical consequence of the situation arising in quantum mechanics. For it is only because *both* the object and the apparatus can be considered classically, in classical physics, that the influence, in general always present, of measuring devices on measured processes may be suspended. According to Bohr,

> This necessity of discriminating in each experimental arrangement between those parts of the physical system considered which are to be treated as measuring instruments and those which constitute the objects under investigation may indeed be said to form a *principal distinction between classical and quantum-mechanical descriptions of physical phenomena*. It is true that the place within each measuring procedure where this discrimination is made is in both cases largely a matter of convenience. While, however, in classical physics the distinction between object and measuring agencies does not entail any difference in the character of the description of the phenomena concerned, its fundamental importance in quantum theory . . . has its root in the indispensable use of classical concepts in the interpretation of all proper measurements, even though the classical theories do not suffice in accounting for the new types of regularities with which we are concerned in atomic physics.[50]

The influence of these interactions thus cannot be disregarded, while taking them into account enables one to connect the mathematical formalism of quantum mechanics with the experimental conditions under which quantum data are produced. By the same token, however, these data become connected to the mathematical (e.g., Hamiltonian) formalism of classical physics that describes measuring instruments in use. It is clear from Bohr's analysis that the functioning of apparatus according to this formalism, especially the exchange of momentum and energy and the corresponding conservation laws (applicable in both classical and quantum physics), plays a crucial role. Even simple measurements of length entail a great deal of classical physics; and, of course, such measurements are never simple in view of the theory of relativity (to which Bohr related his argument on several occasions). It follows that there exists a certain correlation or, one might say, extending Bohr's correspondence principle, a certain interactive correspondence between the *quantum physics* of the measured object and the *classical physics* of the measuring apparatus. The quantum formalism describing quantum systems and the classical formalism describing the apparatus must be both compatible and interactive, must (to take advantage of the double meaning of the word) *correspond* to — and *with* — each other. In view of quantum indeterminacy, the messages involved can never be fully decipherable by classical means, and a "language" that is fully quantum may never be accessible to us at all.

In fact, it appears (assuming that Bohr's argument, based on the classical description of measuring instruments, is correct) that if one assigns classical-like states, a classical wholeness or any classical-like picture, to quantum objects and, by implication, if one claims a *realist* interpretation for the data described — in a *nonrealist* way — by quantum mechanics, then one will no longer be able to maintain a classical description of the measuring instruments. Classical macrophysics and quantum microphysics become intertwined in a very interesting and asymmetrical way: attempts to introduce a more classical or classical-like description at the microlevel appear to break down a classical description at the macrolevel, while nonclassical quantum micro-description is fully consistent and sustains classical macro-description. This is a startling result. Inspired by the ideas of Einstein, Louis de Broglie, David Bohm, J. S. Bell, and others, theories that aim to restore a classical-like picture to quantum data would lead to a violation of classical physics for measuring instruments. Quantum and classical physics depend on and support each other, at least in quantum measurement situations.[51]

These conclusions would not have surprised Bohr, given his interpretation's fundamental basis on the irreducible interactions between a — *quantum* — object and a — *classical* — measuring apparatus. In view of these interactions, an unambiguous meaning can be attributed to the expression "physical reality" in quantum situations only to a very limited extent, if any. Quantum mechanics instead represents or (to employ the term that Bohr would more likely use) symbolizes something else — namely, the structure of the interaction between quantum objects, which would disallow a realist interpretation, and classical measuring instruments, whose physics would allow a realist interpretation. It connects the possibility of realism and classical conceptuality at the macrolevel to their suspension at the microlevel.

This is one of Bohr's greatest points, which was missed by Einstein, who thought quantum mechanics to be a kind of trick, allowing one to explain the data and to make correct predictions, but having no coherent physical or philosophical meaning. Once Bohr's interpretation is understood, however, it is easy to see why he viewed quantum mechanics and complementarity as neither tricky nor mystical, but as "a rational utilization of all possibilities of unambiguous interpretation of measurements, compatible with the finite and uncontrollable interaction between the objects and the measuring instruments in the field of quantum theory."[52]

Beyond its extraordinary contribution to physics, Bohr's interpretation of these relationships between classical and quantum physics has major philosophical significance, especially if one accepts the understanding crucial to Bruno Latour and other recent authors. On the one hand, our experimental equipment embodies our, often hidden, theoretical (or philosophical and ideological) assumptions. On the other hand, reciprocally, our theories and other idealizations (including our philosophical and ideological ones) are irreducibly equipment-laden —technological—leading to a deconstruction of the classical understanding of both theory and technology, and to a recomprehension of both concepts through a more interactive matrix. This understanding can again be linked to Heidegger's insight concerning the jointly mathematical-experimental character of modern science, but it also extends and radicalizes Heidegger's understanding of the situation. Bohr's interpretation can be construed as an exposure of this interactive reciprocity between formalism and experimental technology—a kind of deconstruction of formalism and idealization—and, reciprocally, of the concept of technology (if uncritically understood) *from within* physics itself. For, although one cannot in general suspend the interactions at issue in view of the broader conditions (technological, ideological, social, cultural, and other) affecting any scientific field, classical physics allows one to disregard them, at least within the limits of its functioning as a physical-mathematical field, and, by the same token, becomes permissive of a realist interpretation. By contrast, from the quantum postulate and uncertainty relations on to the most recent experiments, all quantum data available to us make it impossible to disregard the interaction between formalism and experimental technology and, as a corollary, to suspend the possibility of a realist interpretation. Bohr's argument shows that one cannot maintain a realist or any classical-like interpretation without coming into conflict with quantum data, while complementarity offers a consistent and comprehensive account of these data. The consistency and predictive power of quantum mechanics was recognized by Einstein, even though he viewed quantum mechanics as deficient on philosophical grounds and believed that a more classical-like theory would eventually emerge.[53]

Complementarity does not abandon classical representations or — and *as* — idealizations. It utilizes them in order to approach that (radical alterity) which is classically (or, indeed, by any means) unrepresentable and unidealizable. This is what our classical equipment (at whatever level and in whatever sense of the term) appears to tell us. There appears to be an irreducible interdependence of classical and postclassical frameworks in the mathematical-scientific and, possibly, the philosophical fields. This interdependence does not imply a uniform symmetry but specific connections and disconnections, or symmetries and asymmetries, between different classical and postclassical configurations.

Quantum physics has major implications for our understanding of the nature of mathematical representation and idealization, as the relationships between continuous and discontinuous mathematical objects demonstrate. These relationships are not only crucial to quantum physics itself, but they define mathematical idealization, particularly if coupled with the question of infinity and finitude. This coupling is irreducible because continuous mathematical objects, especially the continuum of (so-called) *real* numbers, are fundamental to considerations concerning the nature and structure of mathematical infinitude.[54] The complementarity of continuity and discontinuity is, as we have seen, central to quantum mechanics. It was crucial to Bohr's thinking, beginning with his early encounters with Harald Høffding's philosophy and Bernhard Riemann's mathematical ideas. Høffding placed the relationships between continuity and discontinuity at the center of both his own system and his view of the history of philosophy.[55] His doing so was understandable. These relationships have been of paramount significance in the history of philosophy, mathematics, and physics alike. As Bohr observed, "The more precise formulation of the content of the quantum theory appears . . . to be extremely difficult when it is remembered that all concepts of previous theories rest on pictures which demand the possibility of *continuous* variation."[56] He remarked elsewhere, "Every notion, or rather every word, we use is based on the idea of continuity, and becomes ambiguous as this idea fails."[57] It is not possible to explore here the role of this problematic in the history of mathematics or in the debate concerning the foundations of mathematics, notably, the relationships between and the relative primacy of two respective idealizations (the concept of number and the continuous concepts of geometry), or their possibility, to begin with, as well as their relative primacy in different approaches to the foundations of mathematics, and so forth. Here, I can only address the implications of quantum physics for this problematic.

As we have seen, in quantum mechanics all measurements are performed classically and all measuring instruments are described in terms of classical physics, even though the interactions between quantum objects and measuring instruments are themselves quantum. The quantum character of these interactions may mean that they are strictly finite or, in view of the indeterminacy involved, complementarily continuous and discontinuous, finite and infinite, or, as subject to the regime of radical alterity, finally neither. This situation has crucial implications for the question of the very "existence" of the mathematical objects involved and of all mathematical objects, especially numbers, through which all measurements in physics are performed. For what would it mean to say that such mathematical objects as real, or integer, numbers exist? How could their existence be established or verified? In view of the scales involved, quantum physics is bound to play a major role in establishing the limits of the material realization of mathematical objects — that is, their correspondence to any form of matter, or rather to whatever can be *reciprocally* constructed as such (mathematically, technologically, philosophically, or still otherwise). This reciprocity is, as we have seen, irreducible.

Given the preceding discussion, one would hesitate to speak of the physical reality of numbers, or to reduce the functioning of any mathematical object — infinite or finite, continuous or discontinuous, or, again, conceivably neither — to the possibility of its realizability, its physical embodiment or construction. It is especially difficult to do so in view of Bohr's interactive interpretation of quantum mechanical formalism, in which infinite and, most likely, (directly) physically unrealizable mathematical theory "represents" the interactions between quantum objects and measuring instruments. Perhaps irreducibly involving quantum physics, however, the question of the physical embodiment of mathematical objects is important. As Roger Penrose notes:

> The relationship between the abstractly defined "real" numbers and physical quantities is not as clear-cut as one might imagine. Real numbers refer to a *mathematical idealization* rather than to any actual physically objective quantity. The system of real numbers has the property, for example, that between any two of them, no matter how close, there lies a third. It is not at all clear that physical

distances or times can realistically be said to have this property. If we continue to divide up the physical distance between two points, we would eventually reach scales so small that the very concept of distance, in the ordinary sense, could cease to have meaning. . . .

The real number system is chosen in physics for its *mathematical* utility, simplicity, and elegance, together with the fact that it accords, over a very wide range, with the physical concepts of distance and time. It is *not* chosen because it is known to agree with these physical concepts over *all* ranges. One might well anticipate that there is not such accord at very tiny scales of distance and time. . . . We should at least be a little suspicious that there might eventually be a difficulty of fundamental principle for distances on the tiniest scale. As it turns out, Nature is remarkably kind to us, and it appears that the same real numbers that we have grown used to for the description of things at an everyday scale or larger retain their usefulness on scales much smaller than atoms — certainly down to less than one-hundredth of the "classical" diameter of a sub-atomic particle, say electron or proton — and seemingly down to the "quantum gravity scale," twenty orders of magnitude smaller than such a particle! . . . The appropriateness of the real number system is not often questioned, in fact. Why is there so much confidence in these numbers for the accurate description in physics, when our initial experience of the relevance of such numbers lies in a comparatively limited range? This confidence — perhaps misplaced — must rest (although this fact is not often recognized) on the logical elegance, consistency, and mathematical power of the real number system, together with a belief in the profound mathematical harmony of Nature.[58]

The extrapolation of real numbers to new scales is predictable, given the history of their role in physics; and they may prove to be effective to the limits of what is available to us. Such a possibility, however, whether for real numbers or for any other mathematical object, is far from assured. Their functioning can be understood in postclassical terms, which understanding would be very different from that of Penrose himself. (Penrose's comments here need in fact to be further nuanced in mathematical and physical terms as well.) His views lead Penrose to suggest stronger connections between physical reality — or "the profound mathematical harmony of Nature" — and complex numbers, whose significance for quantum physics is elegantly discussed by him. Penrose's Platonist claims concerning both mathematical and physical reality are problematic from the perspective outlined here. He does, however, correctly point out that complex numbers, like any mathematical object or model, are no less — and, I would add, no more — *real* than any other. Naturally, not all models are equally effective or acceptable from a physical or mathematical standpoint. Given that our physical theories are incomplete, it is conceivable that more complex and yet unheard of mathematical, physical, or metaphorical models will be more effective; and, again, Bohr's interpretation introduced new dimensions into this question. It is also possible that a return to a more classical picture will eventually result, as Einstein hoped.

From the perspective suggested here, all mathematical objects must be seen as idealizations — be they infinite or finite, continuous or discontinuous, or still other — irreducible to any classical reality (or classical ideality), however different they may be in their functioning and in their relation to the local constraints under which we operate in mathematics, physics, or elsewhere. It is by no means clear that physics at the quantum gravity scale (where gravitation must be taken into account in considering quantum interactions) will have to rely more on finite (or discontinuous) than on infinite (or continuous) models, for example, if string theories or other recent theories hold. One might argue that these models are no less real — that is, no more and no less fictional — than real numbers or integer numbers, which may appear to be more physically, materially realizable. While they may be more materially constructible at a certain level, finite mathematical objects are no less idealizations than infinite ones. We might be dealing with "infinite" idealizations in either sense of the phrase — a bottomless interminability of idealization and the irreducibility of idealizations in terms of infinite models. At the limit, matter (as radical alterity) may be neither finite nor infinite (or, again, neither material nor mental). We may eventually arrive at mathematical objects that are likewise neither infinite nor finite, and we may even construct material realizations of such models as effects of the radical alterity-efficacity considered here; and modern — and postmodern — mathematics *continuously* complicates the relative status of continuous and discontinuous mathematical objects and the difference between their mathematics. Ultimately, we may have no more (or no less) intuition of finitude or discontinuity than of infinity and continuity, particularly as regards geometrical objects, such as continuous intervals. It is worth noting that most representations and idealizations of mental, particularly conscious, processes are continuous. This apparent continuity has been crucial to the history of philosophy as the metaphysics of presence, from or before Plato and Aristotle on, and also, accordingly, to the deconstruction of this metaphysics by Nietzsche and by Derrida. The metaphysics of presence is often the metaphysics of the continuum. The concept of mathematical continuity or infinity is part of this history, and the proximity between philosophy and mathematics (or physics) in this regard cannot be ignored. One must, however, equally respect the differences in definition and functioning of these concepts in different fields. Infinitist and continuous mathematics and physics can be used within postclassical regimes, as in Bohr's, while finitist models in mathematics or elsewhere can be and have been complicit with ontotheology and the metaphysics of presence, to return to Derrida's terms. For the latter can also take finitist (or materialist) forms, not only infinitist (or idealist) ones, however prevalent infinitism and idealism may have been in this respect. The one, that defining finite number, and the infinite have conditioned and mutually defined each other ever since Parmenides, at least. Given our own perceptual and conceptual enclosure(s), their unconditional dissociation — the final divorce of the finite and the infinite — may not be conceptually possible, even if strictly finite (or, conversely, strictly infinite) models become more desirable in mathematics, physics, or elsewhere. All such models (infinite, finite, or complementary) and the differences, complicities, and interactions between them must be accounted for in terms of postclassical efficacities.

Real numbers or other mathematical models are idealizations and perhaps, at the limit, are materially unconstructible in view of the quantum character of matter. Real numbers form the mathematical basis of measurement in physics and offer frequently indispensable approximations. All actual measurements so far remain within relatively restrictive limits, although mainly for technological reasons. Modern computers can generate far-reaching approximations of non-rational numbers and of the continuum. Naturally, these approximations fall *infinitely* short of infinity. It need not follow, however, that (the radical alterity of) matter can be mapped by rational numbers within certain limits, albeit quite large, and is thus finite. It is true that our measuring instruments have not encountered, or rather registered, and perhaps can never register any infinities. Would one be able to claim, however, that any given number — say, "one" — is fully materially realizable? Such a claim would already be problematic in view of the quantum character (which, at least for the moment, appears inescapable) of all processes involved in such realizations. Many other problems entailed by this claim should be apparent in view of the preceding discussion. Modern physics depends on both finite and infinite mathematics. It would not be possible without infinitist models and calculations based on them. The very numerical limits (both micro and macro) at issue in modern physics are established through the application of such models as Riemann's manifolds, Hilbert's spaces, and so forth. Experimental procedures and devices themselves, and indeed all modern technology, would be impossible without them, however finite the results of actual measurements are. Physics so far has used infinite models with extraordinary effectiveness, specifically in constructing a comprehensive and coherent understanding of quantum mechanics and its relation to classical physics. It is true that we do not have a fully consistent physical theory at the quantum level. Even leaving aside quantum gravity or the interaction between nuclear and electroweak forces, the necessity of using so-called renormalization in order to circumvent infinite magnitudes deprives even quantum electrodynamics (experimentally the best confirmed theory so far) of full consistency. Obviously, in a fully finite model there would be no infinities of any kind. Such models, however, may not be effective or even possible, given the experimental data we have; and again, especially in view of Bohr's interactive interpretation, these data may be unobtainable without infinitist and continuous models, and the experimental technologies that depend on them. All finite limits at issue are calculated by means of infinitist mathematics; and it is highly doubtful that these numbers may be arrived at otherwise. Important considerations are also based on numbers that are likely to remain materially unrealizable. They are larger than any physical medium available to us — such as the number of elementary particles in the universe.[59] Finally, some string theories, while infinite, are free of problematic infinities and need not depend on renormalization.

We may be dealing with forever unrealizable complexities — or simplicities (for these concepts, or again all concepts that are or ever will be available to us, may not be applicable). There may never be a mathematically consistent model of matter, or a model for and a way to conceive of its radical alterity. Nature may have no mathematical harmony and, in the end, may not be kind to us. Supplementing (in either sense) Eugene P. Wigner's view with complementarity, we may say that in addition to "the unreasonable effectiveness of mathematics in the natural sciences," we face its reasonable ineffectiveness there, although its effectiveness may not be as unreasonable as Wigner seems to suggest.[60] Matter (as radical alterity) may be neither finite nor infinite, nor may it conform to any idealization we might develop. That which would be neither finite nor infinite already appears to be (at least so far) inconceivable (rather than merely unvisualizable or materially unconstructible, as many mathematical objects are), which is not to say that such a conception could not eventually be developed.[61]

The framework of radical alterity suggested here does not imply that nothing can be inferred about this alterity or that new knowledge, theoretical or otherwise, cannot be developed, including knowledge concerning new effects of this alterity. This is one reason for insisting that this radical alterity cannot be seen as absolute. By definition, however, all inferences from and knowledge about this alterity "itself" must be oblique or indirect, since we can only know its effects, whether more immediate or more mediated, even though and because the very relationship between causes and effects becomes refigured as a result. What is at stake in this analysis is what is *un*realizable and *un*representable, including what is unrepresentable as the *absolutely* unrepresentable. At the same time, however, what is unrealizable or unrepresentable is not without productive effects. Uncertainty relations have been and continue to be a key constructive instrument of modern physics, although the latter is, of course, not reducible to indeterminacy. It is conceivable that the so-called Planck scale (length about 10^{-33} cm, time about 10^{-44} seconds, and temperature about 10^{-32} degrees) represents the limit of the mathematical idealization of physical processes, as they are understood and idealized now, or the limit beyond which our mathematico-technological enclosure prevents us from reaching. It is equally conceivable, however, that the whole situation will be refigured, within the same or new limits, and all enclosures involved — mathematical, technological, or conceptual — will be transformed. In physics and mathematics alike we may forever face not "a central unresolved difficulty," but an "extremely large number of inaccessible conjectures," as Robert P. Langlands says, and the configuration may not form "a coherent whole," as Langlands hopes.[62] For among the effects of the radical alterity at issue is a heterogeneity which may never allow a full synthesis. We may have to use an unheard of mathematics, combining new and old models in unprecedented ways. We may need idealizations that combine more diverse and more radically incompatible models, old and new, than does present-day physics. We may be entering, indeed already inhabiting, increasingly complex landscapes that enlarge the space of the known and the unknown alike, which is perhaps the only sense of progress one can still think of.

NOTES

1   Heidegger, rightly, sees the mathematical and experimental aspects of all modern (i.e., post-Galilean) science as fundamentally interrelated. See his *What is a Thing?* trans. W. B. Barton, Jr., and Vera Deutsch (South Bend, IN, 1967), 93.

2   Friedrich Nietzsche, *Philosophy in the Tragic Age of the Greeks*, trans. Marianne Cowan (Chicago, 1962), 117, 112; translation modified.

3   See Derrida's elaborations in *Margins of Philosophy*, trans. Alan Bass (Chicago, 1982), 11, 26–27, 65–67; and *Positions*, trans. Alan Bass (Chicago, 1981), 42–45; see also Derrida's remarks on Nietzsche in *Of Grammatology*, trans. Gayatri C. Spivak (Baltimore, 1976), 19. I consider the notion of (en)closure in detail in *Complementarity: Anti-Epistemology after Bohr and Derrida* (Durham, 1994), 225–69.

4   See "Différance," in Derrida, *Margins*, 7, 11, 26–27, where he refers specifically to Saussure's, and implicitly to Hegel's and Husserl's, among others', classical concepts of concepts.

5   Concentrating on Bohr, I will not be able to discuss these figures here, and my argument is intended to be independent of their texts. Their work, however, is crucial to this essay. I have considered the connections between these thinkers and quantum mechanics elsewhere; see Arkady Plotnitsky, *In the Shadow of Hegel: Complementarity, History and the Unconscious* (Gainesville, 1993); and, especially, *Complementarity*. Several portions of the present article draw substantially on the argument developed in the latter study. My presentation here is, however, independent and develops the problematics at issue in a somewhat different direction.

6   See, for example, Abraham Pais's excellent account in *"Subtle is the Lord . . .": The Science and the Life of Albert Einstein* (Oxford, 1982). See also Thomas S. Kuhn, *Black-Body Theory and the Quantum Discontinuity* (New York, 1978).

7   Pais, *"Subtle is the Lord,"* 369.

8   Abraham Pais, *Inward Bound: Of Matter and Forces in the Physical World* (Oxford, 1986), 134; Albert Einstein, *Out of My Later Years* (New York, 1950), 229; Niels Bohr, "Quantum Physics and Philosophy: Causality and Complementarity," in *The Philosophical Writings of Niels Bohr*, 3 vols. (Woodbridge, CT, 1987), 3: 2. Both are cited in Abraham Pais, *Niels Bohr's Times, in Physics, Philosophy, and Polity* (New York, 1991), 87.

9   Pais, *"Subtle is the Lord,"* 371.

10   Ibid., 404. (Pais combined statements made by Einstein in two 1909 papers.)

11   Ibid.

12   Bohr, *Philosophical Writings*, 1: 8.

13   Quantum mechanics itself was introduced earlier by Heisenberg and Schrödinger as, respectively, matrix mechanics and wave mechanics. These two formulations of quantum mechanics were found to be equivalent.

14   Bohr, *Philosophical Writings*, 1: 54–55.

15   The complementarity of time and energy involves additional complications, which I shall bypass here.

16   Bohr, *Philosophical Writings*, 1: 53.

17   Ibid., 1: 53–54; my emphases.

18   Ibid., 1: 56–57; my emphases.

19   Werner Heisenberg, *Physics and Beyond* (New York, 1971), 210.

20 Derrida, *Of Grammatology*, 50; translation modified. See also the earlier discussion of Nietzsche (19).

21 Werner Heisenberg, "Remarks on the Origin of the Relations of Uncertainty," in *The Uncertainty Principle and the Foundation of Quantum Mechanics*, ed. William C. Price and Seymour S. Chissick (London, 1977), 8.

22 Jacques Derrida, *Speech and Phenomena and Other Essays on Husserl's Theory of Science*, trans. David B. Allison (Evanston, 1973), 89. On Derrida's notion of trace, see especially such major earlier texts as *Of Grammatology; Positions;* and "Différance"; and "Freud and the Scene of Writing," *Writing and Difference*, trans. Alan Bass (Chicago, 1979), 196–231.

23 See Friedrich Nietzsche, "The Will to Power as Knowledge," *The Will to Power*, trans. Walter Kaufmann and R. J. Hollingdale (New York, 1968), 261–331 (one among many of Nietzsche's works that could be cited in this context). My term "efficacity" here is designed to indicate and to emphasize this difference from classical causality. I have developed this concept in more detail in *Complementarity* (see esp. 72).

24 Derrida, *Positions*, 26.

25 For these analyses, see especially Derrida, *Of Grammatology; Speech and Phenomena;* "Freud and the Scene of Writing"; and "The Linguistic Circle of Geneva" and "The Supplement of Copula," in *Margins of Philosophy*, 137–53, 175–205.

26 Complementarity is not fully analogous to the Derridean economy. I develop this argument in more detail in *Complementarity*, 191–269. One of the most significant applications of complementarity was Max Delbrück's work in — indeed in many ways *founding* — molecular biology. Delbrück's thinking was inspired both by complementarity as a general concept and by Bohr's specific ideas concerning complementary thinking in biology. Before he went on to make his groundbreaking contribution in biology, which eventually brought him a Nobel Prize, Delbrück worked at Bohr's institute in Copenhagen. See Max Delbrück, "A Physicist Looks at Biology," *Transactions of the Connecticut Academy of Arts and Sciences* 38 (December 1949): 173–90.

27 Niels Bohr, "Can Quantum-Mechanical Description of Physical Reality Be Considered Complete?" in *Quantum Theory and Measurement*, ed. John Archibald Wheeler and Wojciech H. Zurek (Princeton, 1983), 148.

28 John Archibald Wheeler, "Law without Law," in Wheeler and Zurek, eds., *Quantum Theory and Measurement*, 183.

29 Ibid., 184–85.

30 Ibid., 183–84.

31 Ibid., 189.

32 Derrida, "Différance," 21.

33 See Wheeler, "Law without Law," 190–92.

34 See Bohr, *Philosophical Writings*, 2: 57.

35 See Derrida, *Of Grammatology*, 66–67.

36 Letter to Max Born, 2 March 1953 (Niels Bohr Archive, Copenhagen; Manuscripts and Scientific Correspondence); cited by Henry J. Folse, *The Philosophy of Niels Bohr: The Framework of Complementarity* (Amsterdam, 1985), 248.

37 Bohr, *Philosophical Writings*, 2: 51, 63–64.

38 Ibid., 1: 2.

39 Bohr, "Quantum-Mechanical Description," 146.

40 See Bohr, *Philosophical Writings*, 1: 11; and 2: 25–26, 72–74.

41   Ibid., 1: 54, 66–68.

42   Derrida, *Positions*, 64.

43   Derrida, "Différance," 21.

44   Cf. Bohr, *Philosophical Writings*, 1: 9. See also Andrew Pickering's article on Hamilton's other major discovery or "construction" — quaternions — "Concepts and the Mangle of Practice: Constructing Quaternions," in this volume. At the core of Hamilton's work on quaternions is the question of relationships and, specifically, of the structural isomorphism (i.e., the correspondence between the relations and operations, in addition to the elements themselves) of algebraic and geometrical objects, such as real numbers and the line or complex numbers and the two-dimensional plane. However, fundamental connections to physics, on the one hand, and to philosophy (specifically, the question of reality), on the other, were central to Hamilton's thinking during the genesis of quaternions. In addition, new dimensions of all these questions were suggested by Hamilton's quaternions themselves, as they were by the non-euclidean geometries of Lobachevsky and Riemann, which were introduced at about the same time and subsequently played a crucial role in Einstein's general relativity theory, and by such mathematical objects as the so-called Hilbert spaces used in quantum mechanics. Since Hamilton arrived at the Hamiltonian formalism of classical mechanics the very same year in which he arrived at quaternions, significant and profound connections have emerged, first, between Bohr's and Hamilton's work, and, second, between Hamilton's work on quaternions and subsequent *constructions* (in either sense) in modern and postmodern mathematics and physics, in some of which quaternions themselves played important roles.

45   Noncommutativity, incidentally, also defines Hamilton's quaternions, the first algebraic structure ever introduced that was not subject to the commutative postulate of multiplication.

46   See note 21.

47   See, especially, Bohr, "Quantum-Mechanical Description"; and "Discussions with Einstein," *Philosophical Writings*, 3: 41–47.

48   This argument was crucial to the Bohr-Einstein debate.

49   Bohr, *Philosophical Writings*, 3: 5; my emphases.

50   Bohr, "Quantum-Mechanical Description," 150; my emphases. Bohr's term "phenomena" here refers to the whole interactive configuration comprising both the apparatus and the object. In his earlier writings he used the term "phenomenon" to designate quantum effects observed in classical apparatus. This modification does not affect the argument at issue here.

51   On this point, I am indebted to discussions with Philip J. Siemens and to Anthony J. Leggett's "On the Nature of Research in Condensed-State Physics," *Foundations of Physics* 22 (1992): 221–33, and references there. The search for classical (and specifically deterministic) alternatives to the standard interpretation(s) and to complementarity, in particular, has continued, especially in the wake of David Bohm's "hidden variables" interpretation, introduced in 1952. (For a recent exposition, see David Bohm and Basil J. Hiley, *The Undivided Universe: An Ontological Interpretation of Quantum Mechanics* [New York, 1993].) The appeal of classical thinking and mistrust of postclassical theory have not lost their power, even, as in Bohm's theory, at the expense of sacrificing the main feature, indeed the basis, of Einstein's relativity — the finite (the speed of light) limit on the propagation of action in space, which so far remains a fundamental feature of all known physical, including quantum-

mechanical, processes. It is not surprising, therefore, that Einstein was ambivalent toward Bohm's theory, however much he disliked quantum mechanics. The latter is fully consistent with relativity and all other observed physical data so far, which is a significant point in this context. This is not to say, of course, that one should dogmatically adhere to any interpretation, Bohr's included; and the search for alternatives should not be precluded — How could it be? — even if no new data (contradicting the previous theory) emerges. The question is, however: "In the name of what?" I have considered these issues in detail in *Complementarity*, 149–90.

52 Bohr, "Quantum-Mechanical Description," 148.

53 This is not to say, of course, that quantum mechanics does not in turn depend on extra-physical assumptions, which would introduce further nuances (philosophical, historical, ideological, social, and so forth) into the situation, but would not undermine the point at issue here.

54 Infinity cannot, of course, be subsumed by the continuum in view of the discontinuous concepts of the infinite, beginning with integer numbers or, conversely, higher cardinal numbers (in Cantor's definition). The quantum-mechanical formalism of Hilbert spaces — spaces of an infinite number of dimensions — entails representations involving the power of infinity higher than that of the continuum.

55 See Harald Høffding, *The Problems of Philosophy*, trans. G. M. Fisher (London, 1906), 8. I have considered these aspects of Bohr's thinking in *Complementarity* (121–84), and *In the Shadow of Hegel* (53–95).

56 Bohr, *Philosophical Writings*, 1: 29; my emphasis.

57 Letter to C. G. Darwin, 24 November 1926 (Niels Bohr Archive, Copenhagen).

58 Roger Penrose, *The Emperor's New Mind: Concerning Computers, Minds, and the Laws of Physics* (Oxford, 1989), 86–87.

59 Cf. Penrose's discussion of the probability of "fine-tuning" the physical universe as consistent with the best available physical data, in *The Emperor's New Mind*, 339–45.

60 Wigner made his much-cited comment in "The Unreasonable Effectiveness of Mathematics in the Natural Sciences," *Symmetries and Reflections* (Bloomington, 1967), 222–37. He did not consider the long history of interaction between mathematics and human knowledge and practice, including the natural — *mathematical* — sciences, which could help to explain the situation.

61 The argument made here is thus different from Brian Rotman's critique of mathematical infinitism and his argument for what he calls "realizable mathematics" (or at least "realizable arithmetic"); see his *Ad Infinitum . . . The Ghost in Turing's Machine: Taking God Out of Mathematics and Putting the Body Back In* (Stanford, 1993). Such a mathematics would be fundamentally finitist and discontinuous, and its possibility, functioning, and limits would, equally fundamentally, depend on the physical "realizability" of the mathematical entities one introduces. Rotman appears to see infinitist mathematics as indissociable from infinitist metaphysics or ontotheology, be it mathematical realism or mathematical Platonism, both of which, as well as the role of infinitist metaphysics in the history of mathematics, Rotman is right to criticize. My argument here is that infinitist and continuous mathematics and physics can be used as idealizations within postclassical regimes and can be accounted for in terms of postclassical efficacities, while finitist and discontinuous, or materialist, mathematics and physics may be grounded in the metaphysics of presence, which can be finitist, discontinuous, and materialist no less than infinitist, continuous, and idealist. In view of these considerations, the concept of "realizability" would require

a separate analysis, which cannot be undertaken here. The irreducibility of idealization in all mathematical (at least all topological) constructions, infinite or finite, and the irreducible complexity of the process itself also appear to follow from the "topos theory," one of the most comprehensive mathematical theories ever. It was developed during the last two decades in the wake of Alexandre Grothendieck's work, and it combines, in an extraordinary way, topology (specifically, "sheaf theory") and mathematical logic.

62 Robert P. Langlands, "Representation Theory: Its Rise and Its Role in Number Theory," in *Proceedings of The Gibbs Symposium. Yale University, 1989* (American Mathematical Society Publications, 1990), 209. The article also contains elegant remarks on quantum mechanics (196–97). In this case, "representation theory" refers to the specific mathematical theory or ensemble of theories, which may, however, be interestingly and significantly connected to the general thematics of representation at issue in my discussion.

Is "Is a Precursor of" a Transitive Relation?

E. Roy Weintraub

I N a number of recent papers and a book, I have explored the implications of the idea that the history of economics is neither found nor discovered.[1] I have argued that this history is a creation, a construction by the community of economists.[2] The facts of the history are fluid and mutable, since relevant ideas change as the identity and interests of readers and hearers change and as the community of economists itself becomes concerned with first one, then another, local and contingent problem.

One of the ways in which this construction of the past occurs is through the scholarly disputations of historians of economics. Other changes transpire through the efforts of expositors who seek to explain specialized economic arguments to broader classes of economist-readers. But less well understood is the way in which the past is transformed in the economics community's understanding through the actions of economists: that is, in constructing a bit of economic analysis and placing it in the hands of other members of the community who appreciate the craft and who can be changed in their thinking by the piece of work, individual economists frequently appeal to the past and to shared understandings of that past.

In usual practice, a scholarly paper partakes of a shared vision of what is understood, but that shared vision must be asserted, and, by means of the assertion, alternative understandings are implicitly problematized. This process can be demonstrated by showing how the creation of a model simultaneously creates and is created by history. Of course, this history is a creation of and by language, but it has some dimensions of its own, and the historiographic questions are not entirely addressed in their linguistic dimension. The historiographic element of discussions about methodology is hardly ever remarked upon.

Sometimes "Science" is best examined by following a trail of scientific papers. That is, if we wish to analyze a bit of science, we may read and study a scientific paper as the primary instantiation of scientific practice. If we seek to understand economic model-building, we may be well-advised to begin by reading economics articles which purport to construct models. That is where I too will start—with a paper written several decades ago, but which had a significant impact on its particular subdiscipline, or research program; the paper thus lived on, as it were, in other, subsequent works. Because I have written recently on the history of stability ideas, I choose to consider "On the Stability of the Competitive Equilibrium, I" by Kenneth J. Arrow and Leonid Hurwicz, which appeared in *Econometrica* in 1958. This paper and its companion piece, "On the Stability of the Competitive Equilibrium, II" (coauthored by Arrow, Hurwicz, and the mathematician H. D. Block), are generally taken as together defining the conclusion of a sequence of attempts to establish the stability of the competitive equilibrium, a problem then outstanding in the economic theory literature.[3]

The paper, of nearly thirty pages, has a six-section "Introduction": section 1 sets out the context; section 2, "Dynamic Concepts," sets out the mathematical definitions of equilibrium and stability; section 3, on the adjustment process, suggests two explicit models of market dynamics and frames them in the mathematics of dynamical systems; sections 4 and 5, on excess demands, connect the dynamic process so defined to underlying market behaviors; and section 6 presents a summary of results. It is the following two-paragraph, untitled, introductory remarks to the "Introduction" that I want to consider here:

> A great deal of work has been done recently on what one may call the static aspects of competitive equilibrium, its existence, uniqueness, and optimality. [footnote 2] This work is characterized, in the main, by being based on models whose assumptions are formulated in terms of certain properties of the individual economic units, although in the last analysis it is the nature of the aggregate excess demand functions that determines the properties of equilibria.
>
> With regard to dynamics, especially the stability of equilibrium, much remains to be done. The concept of stability, used already by the nineteenth century economists [footnote 3] in its modern sense, did not receive systematic treatment in the context of economic dynamics until Samuelson's paper of 1941. Samuelson, however, did not fully explore the implications of the assumptions underlying the perfectly competitive model. He (as well as Lange, Metzler, and Morishima) focused attention on the relationship between "true

dynamic stability" and the concept of "stability" as defined by Hicks in *Value and Capital*, rather than on whether under a given set of assumptions stability (in either sense) would prevail or not. [footnote 4] Even though the Hicksian concept does not, in general, coincide with that of "true dynamic stability," it is of considerable interest to us for two reasons: first, as shown by the writers just cited, there are situations where the two concepts are equivalent; second, because the equilibrium whose "stability" Hicks studied is indeed competitive equilibrium. [footnote 5] But again, little is known about conditions under which Hicksian stability prevails. There is thus a gap in this field and our aim is to help fill it. The task consists in constructing a formal dynamic model whose characteristics reflect the nature of the competitive process and in examining its stability properties, given assumptions as to the properties of the individual units or of the aggregate excess demand functions. The results here presented cover certain special classes of cases and many important questions remain open.[4]

The introduction's first sentence notes the previous work on what the authors call "the static aspects" of equilibrium theory, referring (via footnote 2) to results by Wald, Hotelling, Lange, Debreu, Arrow, Arrow and Debreu, McKenzie, Nikaido, Gale, and Uzawa. This sentence defines the immediate context of this paper as a successor to the work done on the "existence, uniqueness, and optimality" of the competitive equilibrium by the named authors. The second sentence characterizes that work as having been concerned with the characteristics of the particular equilibrium which emerges in models based on "individual economic units," or agents (and implicitly on the market behaviors linked to those individual characteristics). The one interesting point to note here, in this history, is that although the papers written by Abraham Wald in the early 1930s are cited, John von Neumann's 1936 paper on equilibrium is not cited, perhaps reflecting the authors' belief that his paper was not part of the sequence of investigations which led to the Arrow-Debreu-McKenzie model and proofs of the existence of a competitive equilibrium. Certainly, the cited papers by Nikaido, Gale, and Uzawa were extensions, and refinements, of the papers on existence (published individually and jointly) by Arrow and Debreu and by McKenzie.

The second paragraph leaves the issues of static equilibrium behind. The first sentence sets up the reader's expectations for this paper by its claim that "with regard to dynamics," especially stability, "much remains to be done." Note, first, that stability is thus presented as a subcategory, a subproblem, in the larger context of dynamics. Second, what is left unargued (and therefore presumed) is that statics is now settled, while stability remains unsettled, leading the reader to expect that the problem of stability is to be taken up and at least partially settled in this paper. (Recall that the title ends with a comma followed by "I," which suggests that at least "II," if not "III," is to follow in due course.)

Although Wald's results had been published only twenty-five years before this 1958 paper, the historicization of Arrow and Hurwicz's argument here is established by their next sentence, which asserts that the concept of stability was already used "in its modern sense" by "nineteenth century economists," with footnote 3 providing references to Augustin Cournot, Léon Walras, and Alfred Marshall.

Turning to these references, we find that Cournot did indeed allude to "no other system of values but the one resulting from these equations being compatible with a state of stable equilibrium."[5] And, in Walras, we read that "such an equilibrium is exactly similar to that of a suspended body of which the centre of gravity lies directly beneath the point of suspension, so if this centre of gravity were displaced from the vertical line beneath the point of suspension, it would automatically return to its original position through the force of gravitation. This equilibrium is, therefore, stable."[6] The last word here is followed by a translator's note of some significance; indeed William Jaffe's note 5 to Walras's "Lesson 7" has been a major source of controversy in the Walras industry, for in it the translator cited Walras's contentious claim to priority in discussing stability, Marshall's counterclaim, and Paul Samuelson's assertion that Walras did not clearly understand the issues at all. In Appendix H of Marshall's *Principles of Economics*, we read about supply and demand curves intersecting at several points and are led to view the various equilibria (intersections) as stable or unstable in the following famous passage: "When demand and supply are in unstable equilibrium then if the scale of production be disturbed ever so little from its equilibrium position, it will move rapidly away to one of its positions of stable equilibrium; as an egg if balanced on one of its ends would at the smallest shake fall down and lie lengthways."[7]

The second half of the Arrow-Hurwicz sentence suggests that these nineteenth-century contributions must be considered unsystematic, for stability did not appear "in the context of economic dynamics until Samuelson's paper of 1941." But that paper too had a defect, since it did not explore the connections, presumably, between the dynamic processes and the ("assumptions underlying") "the perfectly competitive model." Rather, Samuelson, "as well as Lange, Metzler, and Morishima,"

according to Arrow and Hurwicz, had examined "the relationship between 'true dynamic stability' and the concept of 'stability' as defined by
Hicks in *Value and Capital*." The reader is thus led to ask why this particular problem had been attacked by these economists. Why was there
not the more natural examination of the problem of "whether under
a given set of assumptions stability (in either sense) would prevail or
not"? (In Arrow and Hurwicz's footnote [4] here, they do acknowledge
that Samuelson had, in one specific case, done what was being called
for, but they imply that the problem still remained largely unsolved at
that date.[8])

The next sentence begins to outline the paper's central problematic,
stating that John Hicks's concept of stability, imperfect though it appears, "is of considerable interest to us for two reasons." The use of "us"
(the first time the authors have used a personal pronoun) brings the
authors' perspective directly into view for their readers: we are invited
to share in their vision, to see Hicks's contribution, as they themselves
do, as an approach to the true solution of the real problem of establishing stability in an economic model. For although the Hicksian and
"true dynamic stability" ideas sometimes coincide, it is also true that
Hicks established stability, of his type, for an equilibrium that was
"indeed competitive equilibrium." Just so. The footnote here (5) his-
toricizes this assertion with the authors' claim that, in the case of two
goods, this Hicksian problem was "studied by Walras [in the section discussed above,] although without a formal dynamic framework."[9] This
leaves Arrow and Hurwicz's reader to ponder the incompleteness of
the Walrasian analysis, an impression created by the Whiggish use of
"although." "But again, little is known about conditions under which
Hicksian stability prevails" leads the reader to expect that, shortly, the
authors will provide some insight into those conditions. The next sentence promises to satisfy that expectation, making the authors' intentions clearer: "There is thus a gap in this field and our aim is to help
fill it." Again, the first-person pronoun solicits the reader's trust, trust
not only in the validity of the claim that a problematic "gap" exists, but
also in the authors' ability, if not to fill it, then, rather more modestly,
"to help fill it."

The penultimate sentence of this section then defines "the task" of
the paper, on which its success or failure as a paper will ride — namely,
the construction of "a formal dynamic model" that will reflect certain
assumptions, characteristics, and properties. The final sentence reins in
the reader's expectations of that model by restricting the paper's scope
to "certain special classes of cases" and thus invites the reader's appre-

ciation of the task's difficulty, for "many important questions remain open" (and will not be answered in what follows).

One way to begin to reconstruct the genesis of any scientific paper is to encourage the authors themselves to recall just what they were doing when they wrote that paper. In the case of the 1958 stability paper, we are fortunate in that both Arrow and Hurwicz were recently asked quite directly, in interviews, to reconstruct the circumstances and motivations which were responsible for the stability paper's having been written.[10] In response to a question from George Feiwel about the "genesis and essence of your work on stability," Arrow replied as follows:

> That, of course, is more derivative than some of my other work. Stability theory is actually a very old theme, really first discussed by Mill. . . . Samuelson provided a formalization that in fact captured in a general equilibrium sense the intuitive ideas of stability that had been expressed for the most part in a partial equilibrium context. Of course, Walras had made an attempt at a general equilibrium formalization of stability. So, in that sense, the problem had been set forth by Samuelson in his papers in 1941.
>
> In a way, what Leo Hurwicz and I and others were trying to do was essentially to carry out the Samuelson programme and to extend it. Samuelson had a few cases of stability; we tried to extend this in various ways (Metzler had made a very important contribution there). One of the things Leo and I were trying to do was to extend it to global analysis rather than local analysis. Actually, one of the immediate impulses was a somewhat different kind of stability, namely, in programming problems. That again was due to Samuelson. Samuelson had discussed stability, using economic ideas, in the context of linear programming methods. Of course, you are aware that there was an important strand of thought about stability in a socialist system which was in a way between stability of a competitive equilibrium and stability of a programming problem which is a method of optimization. Paul had argued that if you apply straightforward methods in a linear programming system you get cycles.
>
> Leo and I started out by re-examining that question and, in particular, to extend it also to non-linear programming. We got some important results in this simpler context of general equilibrium. Having done that we got interested in the stability concept in the broader context of general equilibrium theory. We used the

standard method in stability theory that had not been used in economics — the so-called Lyapunov method for analyzing global stability.[11]

Arrow's coauthor, Leonid Hurwicz, remembered the nature and development of their collaboration on the stability paper in the following way:

The three areas [of programming, stability, and design of resource-allocation mechanisms] were closely related, at least in their technical aspects: While the first topic dealt with the static aspects of non-linear programming, we then proceeded to the dynamic aspects of linear and non-linear programming. . . . When we shifted to a study of stability of competitive equilibrium, we started out with the thought of applying the techniques we had developed . . . namely the so-called gradient method. (It was Samuelson who had pioneered the use of the gradient method in programming models.) We thought that that method, originally designed for the programming problems would also work in the context of competitive equilibrium and would enable us to determine whether, and under what conditions, competitive equilibrium is or is not stable. So we sort of got into competitive equilibrium theory through programming theory, and we got into dynamics through our initial work on statics. . . . [It] was natural to interpret the dynamics of programming as a certain kind of mechanism for resource allocation. It then became very clear that this kind of interpretation brought us very close to the earlier work of two groups of economists.

One group was composed of people who talked a lot about stability, the main ones being Hicks and Samuelson; the other group consisted of people like Lange and Lerner who were designing a certain kind of economic system. As we started looking at the work of the first group we were struck by the following: On the one hand, you have the work of Hicks which did utilize the concept of competitive equilibrium . . . but whose concepts of stability were not in accordance with the modern notion of what stability means in a dynamic sense. On the other hand, the work of Samuelson was modern in spirit as far as the dynamics was concerned, but it did not exploit fully the specific properties of stability of competitive equilibrium as opposed to any other kind of equilibrium. We noticed there was a gap in the economists' understanding of this problem.[12]

There are a number of interesting features of these interviews. Most important for our understanding of the context of the stability paper is the authors' recollection of a different context than that conveyed by the paper's introductory section. The programming context is missing there, as is the context of the design of resource-allocation mechanisms. Each author of this paper recalls its genesis a bit differently, of course, with Arrow placing a bit less emphasis than Hurwicz on the "design" history, which subject has so engaged Hurwicz's subsequent intellectual attention. Both recall, however, that their interest grew out of a concern with the difficulties of the gradient process as a technique for solving programming problems, raising the following question: Why is the programming context of the 1958 stability paper not present in the paper (and why is the context of the design of resource-allocation mechanisms not present in the paper)? Put another way, why was the problematic of the 1958 paper restricted to the "gap" between Hicks's and Samuelson's treatments of the stability of equilibrium?

It is important to be clear here, for it is not the case that I am accusing Arrow and Hurwicz of inaccuracy or faulty memory, nor am I accusing them of falsifying the historical record in such a way that only an alert historian of economics could set that record straight for posterity. What we have is a conflict between, on the one hand, the paper's 1958 construction of the context for its creation and publication and, on the other hand, a later recollection of a threefold set of ideas which Arrow and Hurwicz contend had organized, and generated, the concerns of that paper in 1958. One way to go about discussing this "problem" is to take an epistemological stand on Truthfulness and The Facts of the Matter. But I have no interest in thinking about the problem of determining which context is the true context. Nor do I believe that this is a problem that requires a solution, if by a solution one means the construction of a metanarrative in which the development of stability theory becomes an exemplum of some larger philosophical point (e.g., that what was really going on was the early stage of a Lakatosian research program, or that there was a Kuhnian paradigm shift or a Laudanian problem-sequence, etc.). Instead, as a historian, I would suggest that it is possible to handle the disparate accounts of what motivated the Arrow-Hurwicz paper by constructing a larger context in which the paper's stated "history" and the authors' recollected "history" can be viewed together and related to a more complex version of "the history" itself. That is, what I wish to suggest is that each of the two versions is appropriate for its own purposes. Different histories are constructed for different purposes and may be equally reasonable for those different purposes (as I shall demonstrate by constructing several other histories that share this feature of problematizing a prior history in the recollection of it, which makes it "false" as well). The historiographic point is "one, two, many histories": pluralistic history, if you prefer. With that as the focus, let me return to the Arrow-Hurwicz paper.

In their introduction, the authors asserted that "the concept of stability ... did not receive systematic treatment in the context of economic dynamics until Samuelson's paper of 1941," but then went on to suggest that Samuelson "did not fully explore the implications of the assumptions underlying the perfectly competitive model."[13] This leaves the reader with the distinct impression that Samuelson had partly explored those implications or, at the very least, that Samuelson had concerned himself with "the perfectly competitive model." This, however, was not the case at all. The structure of Samuelson's paper (which later became chapter 9 of his 1947 *Foundations of Economic Analysis*) makes this clear. An introductory section is followed by a section on comparative statics in which that term is understood in the sense of a formal equilibrium model defined by a set of equations

$$f^i(x_1, \ldots, x_n, a) = 0, i = 1, \ldots n.$$

The next section takes up the Frisch definitions of statics and dynamics and, for a general set of functional equations which include differential-equations systems as a special case, defines what Samuelson calls "perfect stability of the first kind, stability of the first kind in the small, and stability of the second kind." The reference for this material is G. D. Birkhoff's *Dynamical Systems*. I have discussed this connection between Samuelson's development of dynamical theory in economics and his reliance on the works of Birkhoff, Emile Picard, J. Willard Gibbs, A. J. Lotka, Edwin Biddell Wilson, and others elsewhere, so I will not repeat that set of arguments here.[14] Instead, let me point out that the problems which concerned Samuelson were not the problems which concerned Arrow and Hurwicz. Samuelson argued that "in order for the comparative-statics analysis to yield fruitful results, we must first develop a theory of dynamics." To the point, Samuelson's interest in dynamics was specifically associated with his concern to generate comparative-statics propositions, for those propositions, he believed, were the operational theorems of economic theory. We could, as economists, operationalize our theories only if we framed them in terms of *a change in such and such a direction of such and such a parameter would be associated with a rise (or fall) in the equilibrium price (or quantity).* He argued (later dubbing this argument the "Correspondence Principle") that determinate comparative-statics results could be derived from the assumption that the dynamic system was stable in some particular sense.

The remainder of the section on stability and dynamics was devoted to a set of examples to illustrate this connection between "dynamic stability" and comparative statics. For example, his case ends, "If our stability conditions are realized, the problem originally posed is answered. Price must rise when demand increases." His other exempla of this idea are the Marshallian stability conditions, the cobweb, Marshall's trade dynamics, and a combined stock-flow dynamic adjustment in a one-commodity market. Only in another section, "The Stability of Multiple Markets," does Samuelson take up the Hicksian problem. There, he simply frames Hicks's discussion in terms of his own stability definitions and shows that in the symmetric case, for example, *imperfect stability in the Hicks sense necessarily implies perfect stability and conversely.* The next and final section simply goes on to look at the Keynesian system in these same terms of dynamics and comparative statics.

Put most starkly, Samuelson was not concerned with the same set of problems that later concerned Arrow and Hurwicz, and to suggest that he left a "gap" is to reframe Samuelson's contribution as part of a different line of inquiry. Samuelson was not a Walrasian, nor was he in any sense a general equilibrium theorist. Unlike Hicks, Samuelson's reading of Pareto was more influenced by the Harvard "Pareto Circle" of physiologist Lawrence J. Henderson than it was by the general equilibrium basis of Wicksell and Keynes, with which Hicks was concerned. Moreover, Samuelson's view of statics and dynamics was linked more to Gibbs's physical chemistry, as interpreted in social systems by E. B. Wilson, than it was to Walrasian, or even Keynesian, issues. Thus Samuelson did not write on general equilibrium analysis during the flowering of that theory in the late 1940s and early 1950s, concerning himself instead with mathematical economics, notably, the mathematizing of capital theory, public goods theory, and so forth.

Another individual discussed in the tradition called up by Arrow and Hurwicz was Lloyd Metzler (1913–1980). In the area of concern to us, Arrow and Hurwicz cite Metzler's 1945 analysis of the conditions under which Hicksian and "true" stability would be consistent.[15] According to George Horwich and John Pomery, "Metzler showed that if multiple markets are stable for any (relative) speeds of adjustment, then they must satisfy Hicks's concept of perfect stability."[16] What I am interested in suggesting, however, is that this result is less connected to the ideas of the stability of a competitive equilibrium than to a set of issues concerning the nature and meaning of equilibrium and stability, issues which looked back to the 1930s, not ahead to the 1950s. One of the ways to see this is to try to frame some of the ideas in Metzler's 1945 paper in terms of earlier work.

Metzler was an entering graduate student at Harvard in 1937. Paul Samuelson had arrived there two years earlier and was, by the late 1930s, finishing his Ph.D. work and enjoying some scholarly freedom as a member of Harvard's Society of Fellows. His influence on the economics graduate students there then was profound, as can be seen from Metzler's fall 1938 class notes.[17] Enrolled in Joseph Schumpeter's Economics 103 course on theory, Metzler was getting a thorough grounding in the classics of general equilibrium, reading Walras, Pareto, Barone et al., but his class notes also reflect an ongoing exchange between Schumpeter and Samuelson, who was probably attending the lectures. In November, Metzler noted that Schumpeter had an "argument with Samuelson over Harrod's envelope curve"; following an earlier discussion in class, Metzler noted that Samuelson's "original assumption will lead directly to all conditions on indifference curves," in the context of a lecture about which he recorded the fact that "Schumpeter says Walras admits multiple equil. tho Marshall doesn't." (The first comment probably refers to the revealed preference assumptions on which Samuelson was writing in 1938.)

Most interesting is a handwritten note, dated 17 November 1938: "Walras didn't prove solution was unique or even that it existed at all. Wald has shown that a simplified Walrasian system has a unique solution, Mathematical Colloquy [sic] of University of Vienna." This suggests, of course, that Metzler was informed about the most modern and sophisticated work done on the existence of equilibrium in the Walrasian tradition. It also suggests that it was Samuelson, and his concerns, which tended to dominate discussions at Harvard of what was important and significant in recent theoretical work in economics. Schumpeter, with his European connections, had more catholic intellectual tastes than Wilson, Samuelson's unofficial mathematical mentor. And Samuelson was taken up with the same problems that concerned Hicks. Metzler's 1945 paper thus owes much more to Samuelson's influence than it does to the emergent neo-Walrasian program fortified by Wald's and von Neumann's contributions and later linked to the stability literature cited by Arrow and Hurwicz.

The stability literature of the 1940s attempted to navigate among, and stabilize the meanings to be attached to, the differing systemic views on equilibrium and stability associated with Keynesian concerns about unemployment equilibrium and market-clearing failures which would or would not persist over time. This literature reflected ongoing concern with the 1930s problems of dynamics and statics, and of transitions to new equilibria versus returns to preexisting equilibria. Ultimately, of course, the stability literature addressed the metaquestion, "Is the capitalist system stable?" It did not reflect an exploration of the stability of the established equilibrium in a formal model of the competitive system; therefore, it did not entail working out the dynamics of a system of equations whose equilibrium solution had already been formally established. The reconstruction of that 1940s literature as what led to the Arrow and Hurwicz "Competitive Equilibrium" papers of the late 1950s is a Whiggish retelling of the story which relates the interests of the 1940s economists to their future success by means of hindsight and which finds more intellectual kinship among the authors of economics papers than could be reasonably posited except after the fact.

To make this point more vivid, consider a November 1968 letter from Samuelson to Metzler, who had had a brain tumor removed in 1952 and was less able to find the same kind of intellectual energy he had enjoyed in the 1940s. Samuelson, responding to a letter from Metzler about stability, wrote at length on Metzler's contribution to stability theory, placing the 1945 paper in his own Samuelsonian context: "Hicks's (1939) definitions [of stability] are completely in terms of comparative statics. Moreover, in his discussion of comparative statics, both here and in connection with complementarity definitions, his text and Appendix sometimes give slightly different definitions, in which sometimes quantities are the independent variables and sometimes prices." After a discussion of ways to formalize the Hicksian argument, Samuelson continued:

> Now when we come to dynamics, there is the same asymmetry introduced by choosing one of the goods, say the zeroth, as given in the background. That is, we have a different dynamic sequence in time when we use a different *numéraire*. Ideally, we would want our dynamic stability to be such that the system is stable for the actual pattern that takes place in the market place. But who knows exactly what that is? We always work with ideal and slightly artificial models. . . . Hence I write
>
> $$(4'') \; p_i = k_i x_i(p_1, \ldots, p_n), \; i = 1, \ldots nk_i > 0.$$
>
> Now here is where you did Hicks a great service, relating his com-

parative statics conditions to true dynamic conditions. Suppose in $(4'')$, one of the k's gets to be very large, approaching infinity, then by dividing its equation through by that k, say k_n, we find that its statical equation is (practically) always satisfied so as to clear the market. Hence, that equation, and all other equations which have very large k's relative to the rest, can be thought of as (practically) having floating prices that have floated instantaneously to clear their markets.

Now consider the opposite extreme. Suppose some k, say k_2, is "terribly small." Then it is as if that price were frozen; and the same will be true of all such k's that are very small.[18]

After some additional discussion of a matrix condition, Samuelson's letter, written twenty-three years after Metzler's paper had appeared, concludes: "Essentially, there has been no advance on your early theorem." Of course, one of the reasons for this being the case was that Metzler's "early theorem" had been entirely submerged in a tradition in which economic stability theory was linked to the mathematical theory of stability of dissipative dynamical systems. This was one of the causes, and one of the effects, of the Arrow-Hurwicz paper. But the point to be made here is that any claim for Metzler's precursor relationship to Arrow and Hurwicz's concerns does a bit of violence to the context from which Metzler's paper emerged and to the recollection of its context by Samuelson, who had certainly helped to shape its Hicksian problematic.

Let me review the argument to this point:

1. The particular paper under examination, the 1958 Arrow-Hurwicz paper on stability of a competitive equilibrium, historicized its assertions by placing its own arguments in a context in which earlier papers by others were noted as precursors or earlier presentations of the material to be developed in this paper.

2. Although I have only suggested the point, it is reasonably clear that this historicizing of a paper's assertions was not unique to this particular Arrow-Hurwicz paper.

3. The particular historicization asserted in the Arrow-Hurwicz paper can be countered by (a) the retrospective historicization of Kenneth Arrow; (b) the retrospective historicization of Leonid Hurwicz; (c) accounts of Paul Samuelson's concerns with stability ideas by himself and others, as well as the contents of his 1941 paper cited by Arrow and Hurwicz; and (d) accounts by Lloyd Metzler, one of the stability theorists claimed by Arrow and Hurwicz as a precursor.

In order to understand the context of a particular scientific paper or even a mathematical paper, it is necessary to recapture it from the constructed sequence in which it is said, by its authors, to have taken its place. The idea of "precursorness" is not itself inherent; a paper is not published with a footnote stating that "in five years time the results of this paper will be recognized as an imperfect attempt to solve problem X, which is the really important idea, even though the author does not have a clue about what is really going on." To be a precursor of an idea or a solution or a tradition is to be recognized after the fact as having been associated with a line of inquiry which appears in hindsight, from one particular perspective, to have been "on the right track."

What, though, is the meaning to be attached to "on the right track"? If at time t an author A identifies a contribution by B at $t-1$ as having been "on the right track," we see that A is trying to link B with A. Why would this be done? The primary reason to identify oneself with precursors would be to assert claims of "progress." That is, precursorness is often linked to Whiggish histories in which, step by step, the past leads inexorably to the present, where reason marches ahead and progress is reason's measure. The idea that science is an accumulation of reasoned arguments and that science is the exemplar of the growth of knowledge leads naturally to the idea of stages of growth, the passage from less knowledge to more. This construct directs attention to what is common to the growing body, which in turn directs attention to ideational similarities that can be tracked through the progressive sequence of ideas. A precursor, then, is an earlier right idea, where "right" is equivalent to "correct from the present." Science as the exemplar of the march of reason, and economics, as science, leads the Whiggish historian of economics and the typical economic scientist to think in terms of successes and failures, precursors and blind alleys, heroes sung and unsung, and all manner of retrospective gold medals and booby prizes.

Another reason for invoking the idea of a precursor is more rhetorical, more related to the construction of arguments which might be persuasive in the economist's community. That is, A's claim that B is his or her precursor may be an assertion that B has a fame which A can utilize to legitimize his or her own contribution. If I say that my ideas are already present in Keynes's work, then to the degree that I can make such a claim I can buttress my own legitimacy with Keynes's legitimacy. This is an appeal to authority pure and simple, and most claims to have a famous precursor are claims to authority. "It is all set out there in Marshall, or Adam Smith, or Keynes, although in an unformed way," goes the claim of eminent authorization for one's own ideas. Thus the notion that B is my precursor is a rhetorical device which attempts to legitimize certain claims that I might otherwise validate through arguments formulated according to the rules of argument observed within my particular language community.

For economists, the construction of mathematical models, with formalized assumptions and established lines of arguments to conclusions, is but one of several ways of convincing an economist-reader that the assertions are meritorious. Arguments from authority are no less important, even in mathematical economics, than arguments from the data, or from the model, or from the theorem. It is too seldom appreciated that the assertion of a precursor relation is often an argument from authority.

Thus to answer the question posed by the title of this paper, "is a precursor of" is not a transitive relation defined on the set of ideas contained in papers by economists. It is not such a relation because the "set" is not fixed. The ideas of economists are not immutable, even if they are set out in words on a page written or printed in the past, for the ideas themselves are continually recreated. That is, the ideas change in the cascades of representations and re-representations of past work by members of the community of economist-readers. Those economist-scientists choose, for local and contingent reasons, to emphasize or deemphasize certain features of the past in their own work and thus recreate that past by confronting its ideas, and constructions, in their current works. This process never ends. It is one of the ways in which the stock of propositions to which truth-value is assigned changes over time, and one of the ways in which we confer with the larger community of economist-scientists both long dead and yet unborn.

NOTES

This version of the paper has benefited from comments by Neil De Marchi, Bruna Ingrao, and Wade Hands.

1   See E. Roy Weintraub, "Surveying Dynamics," *Journal of Post Keynesian Economics* 13 (1991): 525–44; "Allais, Stability, and Liapunov Theory," *History of Political Economy* 23 (1991): 383–96; "Commentary: Historical Case Studies Are Made, Not Given," in *Post-Popperian Methodology of Economics*, ed. N. De Marchi (Boston, 1992), 355–74; "Introduction," in *Toward a History of Game Theory*, ed. E. Roy Weintraub (Durham, 1992), 3–12; "From Dynamics to Stability," in *Appraising Economic Theories: Studies in the Methodology of Scientific Research Programmes*, ed. Mark Blaug and Neil De Marchi (Cheltenham, 1991), 273–91; "Comment on Heilbroner," in *Economics as Discourse*, ed. W. Samuels (Boston, 1990), 117–28; "Roger Backhouse's Straw Herring," *Methodus* 4 (December 1992): 53–57; "Comment: Thicker Is Better," *Journal of the History of Economic Thought* 14 (1992): 271–76; and *Stabilizing Dynamics: Constructing Economic Knowledge* (New York, 1991).

2   The best general survey of such arguments, as framed by professional historians, can be found in Peter Novick, *That Noble Dream* (Cambridge, 1988). In chapter 15, "The Center Does Not Hold," especially the long section subheaded "Objectivity in Crisis," Novick ranges over the various assaults on "objective" history.

3   Kenneth J. Arrow and Leonid Hurwicz, "On the Stability of the Competitive Equilibrium, I," *Econometrica* 26 (1958): 522–52; and Kenneth J. Arrow, Leonid Hurwicz, and H. D. Block, "On the Stability of the Competitive Equilibrium, II," *Econometrica* 27 (1959): 82–109.

4   Arrow and Hurwicz, "Competitive Equilibrium, I," 522–23.

5   Augustin Cournot, *Researches into the Mathematical Principles of the Theory of Wealth*, trans. N. Bacon (Homewood, IL, 1963 [1897]), 85.

6   Léon Walras, *Elements of Pure Economics*, trans. William Jaffe (London, 1954 [1926]), 109–12.

7   Alfred Marshall, *Principles of Economics*, 8th ed. (New York, 1949 [1890]), 806.

8   Arrow and Hurwicz, "Competitive Equilibrium, I," 548.

9   Ibid.

10 These interviews, conducted by George Feiwel, are the substance of chapters 2 (on Arrow) and 4 (on Hurwicz) of *Arrow and the Ascent of Modern Economic Theory*, ed. George Feiwel (New York, 1987).

11 Ibid., 198–99.

12 Ibid., 258–59.

13 Arrow and Hurwicz, "Competitive Equilibrium, I," 522, referring to Paul A. Samuelson, "The Stability of Equilibrium: Comparative Statics and Dynamics," *Econometrica* 9 (January 1941): 97–120.

14 See Weintraub, *Stabilizing Dynamics*, chapter 3.

15 Lloyd Metzler, "The Stability of Multiple Markets: The Hicks Conditions," *Econometrica* 13 (October 1945): 277–92.

16 George Horwich and John Pomery, "Lloyd Appleton Metzler," in *The New Palgrave: A Dictionary of Economics*, ed. John Eatwell, Murray Milgate, and Peter Newman (London, 1987), 3: 458–61; quotation from 460.

17 These materials are located in the Lloyd Appleton Metzler Papers, Special Collections Department, Duke University Library. Metzler's papers had not been filed or otherwise processed as of March 1994. Metzler's class notes are in file box no. 8.

18 Samuelson to Metzler, 25 November 1968, Metzler Papers, file box no. 13.

Fraud by Numbers: Quantitative Rhetoric

in the Piltdown Forgery Discovery

Malcolm Ashmore

THIS paper is a contribution to two projects: the sociological analysis of debunking and "fraud-busting" in science, and the discursive analysis of quantification rhetoric.[1] It attempts a partial deconstruction of the strong factual status accorded to the result of an exemplary piece of debunking work: the discovery, in 1953–54, of the fraudulence of "Piltdown Man." (For those who are unaware of what John Ziman has famously described as "the only well-known case" of "deliberate, conscious fraud" in academic science, Piltdown Man had been, for the forty years preceding this date, a very famous, though increasingly problematic, paleontological find of some pieces of cranium and jaw which were taken to represent a significantly ancient humanoid; and for the forty years since then, the "same" bones have represented an even more famous, and entirely *un*problematic, case of forgery.[2]) I concentrate here on just one aspect of this debunking work: the ways in which the numerical results of a series of chemical dating techniques (most notably, the "fluorine test") were used to cast doubt on the authenticity of the assemblage. My strategy is to pick as many holes in these results as I can find with the object of weakening (just a little) the unchallengeable character of the Piltdown forgery. *Your* best strategy is to be patient—and to concentrate. The materials I will be working through here are extremely detailed. The rhetorical effect I am after, if you will so oblige me, is similar to that worked up in my protagonists' texts: the overwhelming of resistance through the cumulative effect of "a thousand blows." They were luckier than I, however. Their detailed labors were built into a story with a very dramatic outcome, so dramatic that the details could soon be safely ignored, as indeed many of them have been. So I will own up. There is little drama here. The odd factoid questioned, one or two little puzzles set up. But no denouement, no revelation. Sorry.

Are you still there?

> The great lesson of the Piltdown business for me is that it is un-
> wise to accept current scientific decisions and "proofs" as final,
> irrevocable, and conclusive, no matter how authoritative they may
> sound or look. Always keep in mind the possibility, no matter how
> small it may presently appear, that future evidence and improved
> scientific techniques may alter that proof, conclusion, or decision.[3]

> All the collateral lines of evidence appeared to be mutually confir-
> matory and in complete harmony with each other. . . . So much so,
> indeed, that . . . none of the experts concerned were led to examine
> their own evidence as critically as otherwise they would have done.[4]

These two statements, as post-forgery accounts, intend to refer to, comment on, criticize, and draw moral lessons from the blindness of those
who, prior to the discovery of the forgery, accepted, wrongly, the genu-
ineness of the Piltdown fossils. In short, they are accounts of error.[5] My
aim is to make these statements refer not to the career of Piltdown Man
from 1912 to 1953, but to the career of the Piltdown forgery from 1953
to the present. Doing so will involve going against the advice of a *Nature*
editorial on the occasion of the first announcement of the full extent
of the forgery: "The Piltdown objects have lately been reinvestigated in
such detail, by so many specialists using so many techniques, that there
can scarcely be factual profit from further work upon them."[6] But, then,
the skeptical sociologist of Science has a duty to go against Nature.

The heroes of the story of the discovery of the Piltdown forgery are
Joseph Weiner, of the Department of Anatomy at Oxford University;
Kenneth Oakley, of the British Museum, a pioneer of fluoride relative-
dating techniques; and Wilfred Le Gros Clark, a senior colleague of
Weiner's. Their respective roles in the event were played along these
lines: Weiner was the hypothesizer of the fraud, the initiator of the
investigation, and the revisionist historian; Oakley was the British Museum representative and chief technical tester; and Le Gros Clark was
the establishment figure.[7]

Throughout their public life, the Piltdown bones were subject to two
major interpretations. This interpretative flexibility was occasioned, in
part, by their anatomical makeup (and incompleteness). The first and
most important set of bones, known as "Piltdown 1" (the set from the
first site, a gravel pit at Barkham Manor, Piltdown, Sussex; see Table 1
for a complete list), comprised some pieces of cranium, half a mandible
with two molars attached, and an isolated canine tooth found together

with several other, seemingly prehistoric artifacts and faunal remains. The first interpretation of these finds was that they all belonged to a single individual, popularly dubbed "Piltdown Man." This monistic, single-creature theory was the official line, supported by the British Museum and the scientists closest to the original discovery (and discoverer, Charles Dawson), such as Arthur Smith Woodward and Arthur Keith. It was this version that was celebrated by giving the bones a (single) new genus name — *Eoanthropus dawsoni* (Dawson's dawn man). This theory was reinforced by the 1915 finding of more fragments from a second site (said to be Sheffield Park, about three kilometers from Piltdown), which were taken to represent another individual of the new genus: "Piltdown 2." The original significance of the monistic interpretation was that the bones could thereby represent a possible version of the long-sought-after, and Darwinianly predicted, "missing link" between ape and man. As such, this candidate had a humanlike cranium and an apelike jaw (though with rather humanoid teeth), thus providing potential evidence for human evolution as "brain-led." Archaeological and geological evidence of provenance led to an initial dating of the bones as well over 100,000 years old — thus making Piltdown Man not only the "Earliest Englishman," as it was jingoistically called,[8] but among the very earliest "human" remains known at that time (1912–20). However, from the beginning there were critics (such as David Waterston and Gerrit Miller) who held to the alternative dualistic, or two-creature, theory that the cranium was that of a "human," and the jaw that of an "ape."[9] One feature of the artifacts that legitimized this version was the damage to the jaw, which made its physical articulation with the cranium fragments impossible.

Neither theory fit well with later finds from the 1920s and 1930s, all of which suggested that very early man had more simian cranial features than either version of Piltdown Man, whether represented by the cranial fragments plus the jaw or by the pieces of cranium alone. Thus by the Second World War, Piltdown Man had been shunted into a marginal line in human evolution. All it did was exist: it was no longer significant; it was just embarrassing. And it became even more puzzling as a result of Oakley's (first) fluorine dating tests, carried out in 1949 and published in 1950, which considerably reduced the probable age of the bones from (very roughly) 500,000 to (very approximately) 50,000 years.[10] Crucially, *neither* of the two competing and unresolved theories about the Piltdown remains seemed to make sense given these results. The monistic interpretation suffered if one took the view, as most authorities did, that "no man could have possessed such an apelike form [of jaw] at so late a stage in human evolution."[11] The dualistic theory also faced a severe problem: If the jaw did not belong with the skull, what did it belong with? "The great apes are totally absent from the fossil record of Britain and Europe and are highly unlikely to have inhabited the region during the upheavals of the Ice Age."[12]

Then Weiner came up with his Idea (reputedly on the evening of 30 July 1953).[13] According to his own account, Weiner produced the fraud hypothesis (at this point, suggesting that only the Piltdown 1 *mandible* — and the isolated canine tooth — were fake, planted to make the genuine fossil cranium seem older and more significant) as the only conceivable resolution for these chronic interpretative problems.[14] However, alternative explanations of the puzzle were then available, al-though, post-forgery, they have almost vanished from the record. For example, as Oakley wrote to Keith, "[Robert] Broom recently suggested that *Eoanthropus* may represent an isolated sideline in evolution, in which the brain became large as in *Homo sapiens*, while the jaws evolved in parallel with modern apes. In this case, the later [the] date of *Eoanthropus*, the more 'modern-apelike' one might expect its jaws to be."[15] The contemporary availability of accounts such as this, which save the co-herence of the assemblage even after the first fluorine tests reduced its age, suggests that the forgery hypothesis is somewhat less "necessary" than is usually claimed, most prominently, by Weiner himself.[16]

Another feature of the 1949 fluorine test which would become im-portant only after the forgery hypothesis was in place was the similarity of the results for all of the bones tested (see Table 1, cols. 1 and 3). As post-forgery accounts tend to put it, the results "failed" to differentiate between the age of the cranium and that of the mandible.[17] This was, then, a crucial datum which counted against the idea that the mandible was a modern fake. If Weiner's hypothesis was valid, Oakley's test must be wrong. It is sometimes claimed that Weiner already had doubts about the test prior to getting his Idea: "In 1949, when Oakley had presented his [fluorine test] results, . . . Weiner broached the idea of the mandible being much younger than had been supposed," but "Oakley believed that it was unlikely that a modern specimen could have such a high level of fluorine."[18] However, with the help of Le Gros Clark, Weiner managed to recruit Oakley, who, as a dating expert as well as a British Museum insider, was vital to the success of the project on both a techni-cal and a political level. In putting their case to Oakley, then, Weiner and Le Gros Clark had to persuade him that, among other things, "the first fluorine dates must be incorrect and that all fluorine tests would have to be redone."[19] They were; and the results of this second set of fluorine tests fit well with Weiner's expectations. They showed not only that the bones had generally less fluorine than the first tests had indicated, but also that the mandible and teeth had considerably less fluorine than the cranial fragments (see Table 1, col. 2).

A whole battery of further tests and experiments were carried out which eventually led not merely to the 1953 confirmation of Weiner's original hypothesis that the jaw and teeth had been faked,[20] but to the comprehensive declaration, published in 1955, that every last piece of the Piltdown finds had been forged and planted:

> The mandible has been shown by further anatomical and X-ray evidence to be almost certainly that of an immature orang-utan; that it is entirely Recent has been confirmed by a number of microchemical tests, as well as by the electron-microscope demonstration of organic (collagen) fibres; the black coating on the canine tooth . . . is a paint (probably Vandyke brown); the so-called turbinal bone is shown by its texture not to be a turbinal bone at all, but thin fragments of probably non-human limb-bone; all the associated flint implements have been artifically iron-stained; the bone implement was shaped by a steel knife; the whole of the associated fauna must have been "planted," and it is concluded from radioactivity tests and fluorine analysis that some of the specimens are of foreign origin. The human skull fragments and some of the fossil animal bones are partly replaced by gypsum, the result of their treatment with iron sulphate to produce a colour matching that of the gravel. Not one of the Piltdown finds genuinely came from Piltdown.[21]

These conclusions were instantly and universally accepted (with one exception: the dissenting dentist, Alvan Marston[22]), and the Piltdown forgery has been the hardest of hard facts ever since.

> The development of chemical dating methods makes it possible to settle some of the problems which up to now have been matters of personal opinion. . . . The more such problems can be settled by methods which are *independent of intellectual traditions* the more rapidly our understanding of human evolution will progress.[23]

"Fluorine testing" is a chemical dating technique which was developed in the 1940s as a practical means for the relative dating of anthropological remains from the "neglected" and much earlier work of Middleton and Carnot.[24] Kenneth Oakley, who is usually credited with this development, first used the technique successfully to assess the ages of two

Table 1. Fluorine Contents (%F), Fluorine/Phosphate Ratios ($100F/P_2O_5$), and Nitrogen Contents (%N), as Reported in 1950, 1953, and 1955

British Museum registration number and description	%F 1950	%F 1953/55	$100F/P_2O_5$ 1950	$100F/P_2O_5$ 1953/55	%N 1953/55
Piltdown 1					
Cranium					
Given averages (1953)[a]	[0.2][b]	0.1	—	0.8	1.4
Calculated averages[c]	0.25	0.15	1.3	0.8	0.95
E. 590 L. parietal-frontal[d]	0.1(2)[e]	—	0.5	—	—
E. 590a L. parietal	—	—	—	—	1.9
E. 590b L. frontal	—	0.15	—	0.8	0.3
E. 591 L. temporal	0.4	0.18	2.2	0.8	0.2
E. 592 R. parietal	0.3	0.15	1.8	0.8	1.4
E. 593 Occipital	0.2	—	0.7	—	0.3
Additional fragment[f]	—	0.14	—	0.7	1.6
Jaw, other bones and teeth					
E. 594 Mandible (jaw)	0.2(5)	<0.03	1.0	<0.2	3.9
E. 594 Molars (in jaw)	<0.1	<0.04	0.4	<0.2	4.3
E. 610a Nasal bones[g]	—	0.21	—	1.5	3.8
E. 610b Turbinal bone[h]	—	0.28	—	1.7	1.7
E. 611 Canine tooth	<0.1	<0.03	0.4	<0.2	5.1
Piltdown 2					
Cranium					
E. 646 R. frontal	0.1	0.1	0.8	0.8	1.1
E. 647 Occipital	0.1(2)	0.03	0.6	0.2	0.6
Tooth					
E. 648 Molar	0.4(2)	<0.01	1.3	<0.1	4.2
Controls					
Fresh bone[i]	<0.1	0.03	0.3	0.1	4.1
U. Pleistocene bones 1953[j]	—	0.1	—	0.4	0.7
U. Pleistocene bones 1955 (range of values)[k]		0.14 to 1.3		0.9 to 4.7	0.03 to 3.4
Modern chimp molar	—	<0.06	—	<0.3	3.2

Sources: Compiled from tables in Kenneth Oakley and C. Randall Hoskins, "New Evidence on the Antiquity of Piltdown Man," *Nature*, 11 March 1950, 381; Joseph Weiner, Kenneth Oakley, and Wilfred Le Gros Clark, "The Solution of the Piltdown Problem," *Bulletin of the British Museum of Natural History (Geology)* 2, 3 (1953): 143–44; and Joseph Weiner et al., "Further Contributions to the Solution of the Piltdown Problem," *Bulletin of the British Museum of Natural History (Geology)* 2, 6 (1955): 262, 264–65.

Note: Roughly speaking, the *higher* the number in cols. 1–4, the older the sample; in col. 5, the *lower* the number, the older the sample. Numbers in italics are those from 1955 that do not appear in 1953, both sources having been reports of the same tests undertaken in 1953. Underlined numbers (in cols. 2 and 5) indicate apparent disagreements between the (%N) and (%F) results.

Table 1. Continued

[a] In 1953, the test results for the bones comprising the cranium of Piltdown I (E. 590 to E. 593) were labeled "cranium" and not given individually. Thus these figures are averages, although in the case of both the fluorine (%F) and the nitrogen (%N) content figures, the given values are significantly at variance with my own calculations, based on the results for the constituent bones presented in 1955 (see n. c). The significance is that in both cases the given average, when compared to the calculated average, makes the cranium appear younger.

[b] This figure is the average given for the 1950 cranium results in H. de Vries and Kenneth Oakley, "Radiocarbon Dating of the Piltdown Skull and Jaw," *Nature*, 25 July 1959, 224. Note its variance from my calculation (see n. c).

[c] These figures are my own calculations of the average (mean) values for the (4?; see n. d) constituent cranium bones for which such values were not given in the 1950 and 1955 sources.

[d] Bone E. 590, described in the 1950 source as the "left parieto-frontal," is described in terms of two constituent parts in the 1955 source: E. 590a, the left parietal, and E. 590b, the left frontal. Note that E. 590a was tested only for nitrogen, while E. 590b got the full treatment, with no explanation offered either for this disparity or for the large difference between the (%N) results for these two parts of the "same" bone. Finally, on both types of fluorine measurement, the results for bone E. 590, however it may be divided up, were actually *higher* in the later test than in the earlier one; this bone, that is, looked older in 1955 than it did in 1950. Needless to say, this difference is *not* pointed out in the post-forgery literature.

[e] In col. 1, numbers in parentheses indicate the number of determinations made, if more than one; in these cases, the recorded figure is an average. (See discussion in text of an implication of the five measurements made on the mandible.) There is no mention anywhere of the number of determinations made in the 1953 analyses.

[f] Both types of fluorine test, as reported in the 1955 source, appear to have been done only on this apparently unregistered "additional fragment" and not, as reportedly was the case in 1950, on the occipital bone itself. When comparisons are made between the tests, this difference is ignored, yet to treat these two pieces as the same looks problematic, given the large difference between their (%N) scores.

[g] No mention is made of tests on the nasal bones in either the 1950 or the 1953 source. Note also the contrasting (%F) and (%N) results (see further discussion in text).

[h] No mention is made of tests on the turbinal bone in either the 1950 or the 1953 source. Note also the contrasting (%F) and (%N) results (see further discussion in text).

[i] In the 1950 source, this control sample is described as "fragment of fresh bone from soil"; in the 1953 source, only (%N) results for a sample labeled "fresh bone" are given; and, in the 1955 source, all three test results are reported for a sample labeled "recent; surface, Transvaal."

[j] In the 1953 source, both fluorine test results are described as "minimum F-content of local U. Pleistocene bones," while the nitrogen result is reported for "U. Pleistocene bone (London)."

[k] In the 1955 source, among the list of 14 "bones used for comparison" were 5, described as Upper Pleistocene, which had the range of values given here for each of the 3 tests. Oddly, not one of these 15 (5 × 3) values coincides with any of the 3 values reported in the 1953 source; and, in the case of both fluorine measurements, the figures given in the 1953 source are outside, and lower than, the range reported in the 1955 source. The significance of this is that, between the first announcement of fraud (when the cranium was not yet under suspicion) and the second (when it was), the "fossil barrier," as represented by this control measurement, had been raised. That is, in the 1955 report, "even" the cranium seemed more borderline than it had seemed (even) in 1953.

other contentious British "early-Men": the Galley Hill skeleton, and the Swanscombe skull (or, alternatively, to "test the validity of the fluorine methodology").[25] The tests showed that whereas the Galley Hill remains were much younger than had been thought (by their proponents), the Swanscombe specimen was genuinely ancient. Oakley's next use of the fluorine test was on the Piltdown assemblage.

The test relies on the idea that bones will adsorb fluorine from the environment at a steady rate over time. Thus the amount of fluorine present (in the form of fluorapatite) can serve as an indication of age: the more fluorine, the older. However, the method is one of *relative* dating only, as the amount of fluorine adsorbed depends on the amount in the deposit, which varies from place to place. Therefore, it is best used to assess the relative ages of findings from a single location and thus, for example, to determine if certain bones are later intrusions (the conclusion from the test of the Galley Hill remains). Oakley frequently makes much of the danger of overinterpreting the test's results. Here is an early example of his cautious approach:

> Oakley [in 1943] . . . said the progressive increase in the fluorine content of bones with increasing geological age was directly concerned with the amount of fluorine present in the deposit in which they were found. It was a statistical law, he said, and not applicable to individual specimens. . . . Although *a negative result would not be proof that a bone was recent,* a high fluorine content would be strong evidence of antiquity "in case of doubt arising."[26]

Oakley used this kind of technical argument to correct the "mistake" of directly comparing the numerical results of studies of artifacts from different sites. For example, Marston claimed that because Oakley's fluorine dating of the Galley Hill skeleton and of the Piltdown remains showed them to have similar amounts of fluorine, the latter should be interpreted as of a similar age to the former, which had been estimated to be from the Holocene period (i.e., geologically Recent).[27] Oakley did not agree:

> It is a mistake to suppose that the fluorine content of a fossil bone provides a direct indication of its geological . . . age. In the case of the Piltdown material, fluorine analysis simply showed that . . . the mandible and cranial fragments could not be separated . . . and are contemporaneous with the latest fossils in the gravel. . . . There can be no reason to doubt that *Eoanthropus* is Pleistocene [i.e., earlier than Holocene].[28]

And, when commenting on the later radiocarbon dating results (see below), Oakley is also cautious about the utility and validity of fluorine dating: "Unless one knows the source of a bone, precise relative dating by fluorine content is impossible."[29]

In fact, as we shall see, only when discussing the results of the second Piltdown tests, done under the influence of the forgery hypothesis, does Oakley's technical caution vanish.

> Analytical results of the Oakley-Hoskins 1950 determinations of the Piltdown materials which differ from those of 1953 are presumably to be taken as errors. They do not imply a lack of reliability of the fluorine method for deriving relative ages of bones in the same bed.[30]

It is clearly not sufficient, rhetorically, for immediate post-forgery accounts of the difference between the two sets of fluorine tests to cite the Forgery Hypothesis alone as the reason for the inadequacies of the first set, or indeed as the sole reason for their repetition. Weiner's account, which we have already noted — "the first fluorine dates must be incorrect and . . . all fluorine tests would have to be redone" (see note 19) — was published thirty years later, and posthumously, in an effort to establish that it was Weiner (rather than Oakley) who played the major role in the Discovery. Such bald statements of the test results' incompatibility with the assumption of forgery did not appear at the time. Instead, contemporary accounts of (1) why the second set of fluorine tests had to be done at all, and (2) why their results were judged superior, tend to emphasize *technical* issues. These fall into two camps: improvements in the analytic methodology, and the use of larger and/or heavier samples for analysis.

The notion of *improvement* is a very common trope in scientific discourse; indeed, the famed "progressiveness" of science is in large part textually constituted through its use. (There is a noticeable tension here between evolutionary and revolutionary change: between improvement brought about through continuous modification or through novelty and rupture. How this tension shows itself in this case will be discussed more fully below.) With respect to the second fluorine tests, "improvement" is articulated in the post-forgery literature in two main ways, both of which amount to related yet distinct claims for the "increased accuracy" of the method.[31] The first of these involves the ability to measure smaller amounts of fluorine. This accuracy claim is used as the main technical justification for having the tests redone:

> Improvements in technique have since [1949] led to greater accuracy in estimating small amounts of fluorine, and it therefore seemed worthwhile submitting further samples . . . for analysis.[32]

> In 1953, new samples . . . were submitted to the Department of the Government Chemist, where Mr. C. F. M. Fryd had devised a technique for estimating smaller amounts of fluorine than could be measured in 1949.[33]

But this notion of "increased accuracy" is ambiguous. The ability to measure "smaller" quantities could mean one (or both) of two quite different things. If we think of the space of measurement as a scale, it could mean the ability to distinguish more points on the scale (increased discrimination). Alternatively, it could mean the ability to detect points below the scale's previous lower limit (reduced threshold). It appears that claims for both of these versions of greater accuracy may be warranted if we compare the fluorine content (%F) figures from each test (Table 1, cols. 1 and 2). Increased discrimination could be represented by the change from the one-decimal-place figures of 1950 to the two-decimal-place figures of 1953/55, while the decrease from the lowest 1950 figure (<0.1) to the lowest 1953/55 figure (<0.01) could represent a reduced threshold. Which of these versions is "meant" is unclear from the post hoc accounts, but I would argue that stressing the reduced threshold threatens to undermine the sense of objectivity achieved through the use of a discourse of technical progress. This is because it suggests a prejudgment of the outcome, namely, that there were only very small amounts of fluorine to be found. On the other hand, the increased discrimination version does a great deal of useful rhetorical work. Apart from permitting the crucial possibility of discriminating between the jaw and the cranium (the main aim of the second set of tests), this version's textual representation (the use of figures with two decimal places) manages to *show* the much-vaunted methodological improvement "unarguably."[34] It should be noted, however, that at one point in the 1950 report of the earlier tests, a figure of the "later" order of accuracy is used: "0.05% fluorine."[35] This anomaly apart, it does seem that much of the warrant for the claim of increased accuracy is rhetorically achieved by the striking absence of (the *very* small) hundredths-of-one-percent figures from this text in contrast to their notable presence in the later ones.[36]

The second kind of technical improvement in fluorine testing claimed in the immediate post-forgery texts is the reduction of the method's margin of error by as much as an order of magnitude (from ± 0.2% to ± 0.02%).[37] This form of improvement is used to explain the failure of the earlier tests to distinguish between the Piltdown parts: in 1949, "no appreciable difference was found in the fluorine content of the skull and teeth of Piltdown man; this was due to the margin allowed for experimental error . . . [which] was to have very significant results, though it was not suspected at the time."[38] Spencer, drawing on Weiner's own account, suggests that the margin of error (± 0.2%) attributed to the 1949 test allowed the hypothesis of a modern faked jaw to surmount the obstacle of its published fluorine content (0.2%).[39] Given such a margin, the *real* amount "might well be less than 0.1 per cent — a figure more in line with that of recent bone."[40]

Margins of error are closely linked with the other technical issue mentioned above: sample size and/or weight. Their quantitative relationship appears to be a simple inverse correlation: the larger the sample, the smaller the margin of error. Thus it would appear that the claimed reduction in the 1953 test's margin of error implies a corresponding increase in the size of the samples analyzed; and indeed, this is exactly what is claimed. But, first, we should note that the size of the sample is not the only relevant variable affecting the size of the margin of error. As the forgery's discoverers themselves put it in 1955, "The experimental error in the determination of fluorine obviously depends on the size of the sample *and the amount of fluorine it contains*."[41] Unfortunately, "the amount of fluorine" contained in a sample is precisely what is unknown prior to its determination by the test. This circularity puts all estimations of experimental error radically in doubt (see Table 2).[42]

So what were the margins of error for the two sets of fluorine tests? As can be seen from Table 2, accounts of this issue, like accounts of the related issue of sample size/weight, are highly variable. This variation, however, is significantly patterned. In the case of the 1949 tests, the carefulness and subtlety with which the original published estimates of margins of error are described is not manifested in later texts. Moreover, the later accounts are marked by the virtual disappearance of the lower of the two 1950 estimates and the dominance of the higher estimate. As far as almost all post-forgery accounts are concerned, the 1949 fluorine test had a margin of error of ±0.2 percent.[43] I have set out (in Figure 1) the complete account[44] of the link between the weights of the samples tested and the corresponding experimental errors, exactly as published in 1950, together with my interpretation and conclusions.

My main conclusion from this analysis, then, is that the *highest* margin of error that can legitimately be ascribed to *the majority* of samples is ± 0.1 percent. The naturalization of the figure of ± 0.2 percent — which in the original text functioned as the absolute ceiling — as *the* margin of

Table 2. Margins of Error and Sample Weights in the 1949
and 1953 Fluorine Content Tests

Text	Margin of error (±%)	Sample type and weight (mg)
1949 Test		
1950	"not greater than" 0.1	>5
	"may be [as much as]" 0.2	<5
1953	"accurate only within rather wide limits"	<10
1955a	0.2	—
1955b	—	"too small"
1959	0.1	cranium
	0.2	mandible
Cole	0.2	—
Spencer	0.2	—
1953 Test		
1953	—	"mainly larger samples"
1955a	0.02	—
1955b	—	"larger samples"
1959	0.01	cranium
	— [too small to measure?]	mandible
Cole	—	larger
Spencer	—	—

Sources: Kenneth Oakley and C. Randall Hoskins, "New Evidence on the Antiquity of Piltdown Man," *Nature*, 11 March 1950, 381; Joseph Weiner, Kenneth Oakley, and Wilfred Le Gros Clark, "The Solution of the Piltdown Problem," *Bulletin of the British Museum of Natural History (Geology)* 2, 3 (1953): 143–44; Joseph Weiner et al., "Further Contributions to the Solution of the Piltdown Problem," *Bulletin of the British Museum of Natural History (Geology)* 2, 6 (1955): 262–65; Kenneth Oakley and Joseph Weiner, "Piltdown Man," *American Scientist* 4 (1955): 577; H. de Vries and Kenneth Oakley, "Radiocarbon Dating of the Piltdown Skull and Jaw," *Nature*, 25 July 1959, 224; Sonia Cole, *Counterfeit* (London, 1955), 153–56 (Cole); and Frank Spencer, *Piltdown: A Scientific Forgery* (London, Oxford, and New York, 1990), 137–39 (Spencer).
Note: A dash indicates that no quantification was given.

error for these tests clearly serves the rather obvious rhetorical function of widening the gap between the degrees of accuracy of the two sets of tests. The worse (i.e., the less accurate) the earlier one can be made to look, the greater the perceived accuracy of the later one. As shown in Table 2, the 1949 threshold of accuracy in terms of minimum sample weight suffers a similar fate. In 1950, it is five milligrams; by 1953 (the first announcement of fraud), it has doubled to ten: "The method of

The Account. "Where possible, *at least 20 mgm.* of bone was used for fluorine determination; but *in several cases* it was necessary to rely on samples *of the order of 5 mgm.* The errors of analysis naturally increase as the weight of sample decreases, but it is believed that with sample weights of *5 mgm. and upwards* the error in the adopted values *is not greater than ±0.1* per cent of fluorine. For sample weights *less than 5 mgm.* the error *may be ±0.2* per cent of fluorine."

The Interpretation. In at least some cases, sample weights were greater than 20 mg.; in some proportion of "several cases," sample weights were less than 5 mg.; *therefore* *(though speculatively)*, sample weights varied from c. 4 mg. to c. 25 mg., with the range of "several cases" varying from c. 4 mg. to c. 6 mg., *as shown (diagrammatically)* below:

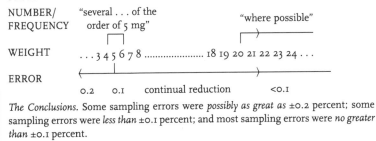

The Conclusions. Some sampling errors were *possibly as great as* ±0.2 percent; some sampling errors were *less than* ±0.1 percent; and most sampling errors were *no greater than* ±0.1 percent.

Figure 1. Interpreting the 1950 account of weights and errors.

analysis used in 1949 was accurate only within rather wide limits when applied to samples weighing less than 10 milligrams."[45] Later texts no longer include any specification of sample weights, apart from occasional comments on their inadequacy: "The amounts we were dealing with [in 1949] were too small to be measured except approximately."[46]

There is created, then, a general impression that the shortcomings of the first fluorine tests were due to the margin of error's being too high because the weight of the samples used was too low. However, there are several problems with this explanation of inadequacy. First, it is far too general. Such an explanation does not account for the fact that, according to post-forgery texts, the first set of tests was only *differentially* inadequate. In particular, the 1949 result for the mandible was taken to be far more wrong than those for the Piltdown I cranial bones. It is this problem that is resolved by de Vries and Oakley's otherwise puzzling decision to split the 1949 (and 1953) test results, together with their associated errors, into those for the cranium and those for the mandible. With respect to the 1949 test, the fluorine-content percentages are given as 0.2 ± 0.1 for the cranium and 0.2 ± 0.2 for the mandible.[47] In terms of Oakley's (own) account from 1950 (see Figure 1), this appears to suggest that the cranium samples weighed "5 mgm. and upwards" while the mandible samples weighed "less than 5 mgm." The use of only a small sample from the mandible in 1949 is again suggested in Weiner and Oakley's 1954 review of the evidence for fraud: "The fluorine content of the mandible . . . was based on analysis of only a few milligrams of material."[48] But this is odd because it is reported in the 1950 text that five determinations were made on mandible samples (see Table 1, col. 1). If this means, as it appears to, that five separate samples were taken from the mandible, then, even if each sample weighed only about five milligrams, the total amount of material collected should have been enough to reduce the margin of error to an acceptable level.[49]

A second problem with the blanket "small sample/large error" explanation for the flaws of the first set of fluorine tests is that it leads to difficulties in explaining variations between the results of the first and second tests that do not lie within the range of even the most pessimistic margin of error ascribed to the first. An obvious example of this problem was noted by Sherwood Washburn in 1953: "The fluorine content of the molar of Piltdown 2 [E. 648; see Table 1] was given as 0.4% in 1950 and now appears as <0.01."[50] This difference entails a margin of error that is nearly double the highest 1950 percentage of ± 0.2 for the smallest samples. Thus it cannot have been due to any recognized experimental error for this test; hence there was an unknown and greater error in 1950, in 1953, or in both. Or perhaps it was just an anomaly — like another problematic datum for the standard account of error we are discussing. How else could supporters of this account explain a case where both the bad and the good tests yielded exactly the same result (see Table 1: Piltdown 2, E. 646, R. frontal[51])?

The final problem I will raise is this one: if the 1949 tests are to be treated as inaccurate due to the imprecision of the method as developed at that time, then all other uses of the method up to around 1953 are also put in doubt — including Oakley's own earlier dating of the Galley Hill and Swanscombe remains as well as his 1951 analysis of the Fontéchavade specimens.[52] But, you will not be surprised to hear, such doubts have not been raised.

The rhetorical process of widening the gap between the two sets of tests is also carried out "from the other end," so to speak. While in the first report of forgery no margin of error is specified at all for the new fluorine tests, Weiner et al. do state that "the new estimations [were] based mainly on larger samples."[53] In a popular account published later, Oakley and Weiner again claim that larger samples were used, although this time the claim is made without the qualification of "mainly."[54] In the second, 1955 forgery report, we *are* given a margin-of-error figure — ± 0.02 percent, which contrasts very neatly with the ± 0.2 percent attributed here to the old tests — but we are not informed of the weights of the samples used. In de Vries and Oakley's 1959 text, the margin of error is suddenly, and without explanation, reduced by half ("0.1 ± 0.01"); but, continuing their practice of splitting the sample types, the authors apply this figure only to the cranium measurements.[55] The fluorine content figure they quote for the mandible ("<0.03"), I believe we are being asked to believe, is so low that no meaningful margin of error can even be computed. If the inverse relation between the sizes of samples and of errors holds here, then one wonders how much of the mandible could have been left after *this* determination!

As I have indicated in Table 2, it is frequently claimed that the samples used in 1953 were larger than the miserable "few milligrams of material" used in 1949.[56] Interestingly, the only contrasting account I have found is at least as authoritative as any of the others. In his own brief contribution to the 1955 report, the government chemist who carried out the tests, C. F. M. Fryd, describes the size of the available samples as "very small . . . of a very few milligrams in weight."[57] Fryd's account also casts doubt on another standard claim that we have already noted: the suggestion that the analytical technique used in 1953 was both different and superior to the one used in 1950. Far from being a radical departure, the 1953 measurements were done "by adaptation of published methods," according to Fryd, who cites Oakley's two earliest studies, *including the 1950 Piltdown work,* as (the only) cases in point![58] Moreover, both sets of tests were conducted in the Department of the Government Chemist, where (the-new-and-improved) Fryd and (the-old-and-outmoded) Hoskins both worked — closely enough, indeed, to have coauthored an article on fluoride dating techniques.[59] It is in the difference between the discoverers' and the technician's accounts of the new tests that the tension between the evolutionary and revolutionary versions of scientific progress, mentioned above, is most clearly felt: where Weiner and Oakley stress novelty, Fryd emphasizes modification.

As can be seen from Table 1, two different measures of the amount of fluorine present were used in both 1949 and 1953. The first measurement, and the only one mentioned outside of the technical literature, is the "straight" fluorine content (%F) determination we have been discussing so far. The second one was a fluorine/phosphate ratio (%F ÷

%P_2O_5 [× 100]). Oakley describes the purpose of this measure in his 1953 review of fossil-dating procedures: "It is now a routine procedure in fluorine dating to check the phosphate content of each sample analyzed and to use the percentage fluorine/phosphate ratio as one basis of comparison, particularly where there has been some alteration or mineral contamination of the bone."[60] One result of the forgery investigations was to establish that most components of the Piltdown assemblage had indeed suffered extensive "alteration or mineral contamination." The investigators took such changes to be further evidence of fakery and were not led by them to question the validity of the percentage-fluorine content results. Yet according to Oakley's own account of the method used in the 1949 tests, this was precisely what they should have done:

> In the case of coarsely porous bone, it is sometimes difficult to obtain a sample which is completely free from silt contamination. The fluorine content of a contaminated portion of a bone will obviously be *misleadingly low*. It was therefore decided to determine the phosphate content of all samples, and to express the fluorine value of each sample as the percentage ratio of fluorine to phosphate (as P_2O_5). This procedure facilitates comparison of the fluorine contents of bones in which there has been variable contamination.[61]

So, one possible explanation for what at the time seemed the surprisingly low fluorine readings presents itself: if any of the bones were found to be contaminated with silt (which does not seem impossible in either a pre- or a post-forgery examination), then their straight fluorine content figures would be "misleadingly low," leading to an interpretation of their ages as misleadingly modern.

In contrast to the 1950 account, the fluorine/phosphate ratio measure is not discussed in either of the two forgery reports, although its results are displayed in their respective tables.[62] And none of the later commentaries that I have read sees fit to mention it.[63] We can ask, then, why the measure that was acknowledged to be problematic in terms of technical adequacy would nevertheless have been the one that was overwhelmingly used in practice. A potentially plausible justification for the evident preference of all concerned to discuss these findings in terms of fluorine-content percentages alone would be that this measure is the "simpler" of the two and thus more readily understood. A more analytically interesting explanation would stress the rhetorical advantage to be gained, post-forgery, from the use of the percentage-fluorine figures rather than the fluorine/phosphate ratios. As can be seen from Table 1, the former are much *lower* than the latter. And because the forgery theory is supported by the bones' appearing to evince the greatest possible degree of youthfulness, the lower the numbers the better.

In 1953, the new attempts to date the Piltdown remains by fluorine-content analyses were supplemented by tests for nitrogen content. "Whereas a fluorine assay reflects the gradual accumulation in bone [and teeth] of an exogenous element, the nitrogen content indicates the progressive loss of organic matter from the bone itself. Accordingly, in fresh or recent bone the nitrogen content is very high while its fluorine content is very low; but with the passage of time this situation is reversed."[64] As is evident in Table 1 (cols. 2, 4, and 5), the nitrogen test results tended to confirm the differential ages of the cranial fragments and of the jaw and teeth, as assessed by the new fluorine measurements. However, one striking anomaly was soon noticed by Robert F. Heizer and Sherburne F. Cook (who are usually credited with developing a reliable method of nitrogen dating): the results for the occipital bone from Piltdown 2 (E. 647).[65] In responding to Sherwood Washburn's summary of the first announcement of forgery,[66] Heizer and Cook claim that his assignment of this bone to the "modern" category, along with the mandible and all of the teeth, ignores the contradictory evidence provided by the percentage-fluorine (0.03) and the percentage-nitrogen (0.6) results. While the fluorine content of the Piltdown 2 occipital bone was far lower than all other pieces of cranium in the collection, and therefore indicated a relatively young age, its nitrogen content was also the lowest of all the pieces sampled, including the "modern" jaw and teeth. And this, of course, indicated a relatively ancient date. As Heizer and Cook comment, this bone could therefore belong to any one of three categories: "(1) a modern bone introduced as a hoax; (2) an archaeological specimen removed from its original site elsewhere and introduced to the Piltdown locality; and (3) an indigenous bone in the Piltdown gravels."[67] In their 1954 review of the evidence for forgery, Weiner and Oakley also noted the anomaly and attempted to resolve it by explaining away the indication of ancientness: "In a bone which may have been exposed on the surface for some time this [low nitrogen content] is no proof of antiquity."[68]

Interestingly, in a later discussion of their findings, Oakley and Weiner chose to make the opposite argument when presented with another apparent contradiction between the fluorine and nitrogen measurements of one of the associated faunal remains: "The . . . hippopotamus molar tooth from Piltdown . . . contains very little fluorine. On the other hand it has lost almost all trace of organic matter which indicates it is *certainly* ancient."[69] Here, then, the lack of nitrogen is cited as evidence for this specimen's being "certainly ancient," while the contrary indication of its lack of fluorine is allowed to pass.

As I have indicated in Table 1, there are two other cases of apparent disagreement between the fluorine and nitrogen test results as reported in 1955. The fluorine test figures for the nasal and the turbinal bones (*E.* 610a and b) are very high (indeed, the highest by far of the whole collection), thus indicating their comparatively great age, while the same bones also yielded relatively high levels of nitrogen, which would indicate their youthfulness. Finally, the set of nitrogen test results for the Piltdown 1 cranium shows considerable variation, while the corresponding sets of fluorine tests do not.

> In 1953–55, the possibility of dating the Piltdown bones absolutely by the radiocarbon method was not seriously considered because it would have involved total destruction of the specimens to provide the minimum quantity of carbon (2 gm.) then demanded by radiocarbon laboratories for a single determination. During the last four years, improvement of technique has made it possible to attempt radiocarbon dating on the basis of much smaller quantities.[70]

Another reason for the failure to employ radiocarbon dating at any time much before this period could be that the greatest age this technique can indicate is approximately 40,000 years. And only in the post-forgery world would it have seemed worthwhile to consider the possibility that the specimens were quite so young. When radiocarbon dating was finally applied in 1959 (after the obligatory "improvement of technique" had occurred), the results were given as follows:

> mandible   500 ± 100 years
> cranium (R. parietal, P1)   620 ± 100 years.[71]

These "absolute" ages are interestingly at variance with the earlier fraud-finding fluorine/nitrogen relative datings. While the latter made the mandible modern and the cranium "fairly ancient,"[72] the radiocarbon dates for the mandible and cranium were so similar as to have a possible eighty-year overlap. As it was precisely the ability of the second set of fluorine tests to *differentiate* the parts of the skull that gave it such high credibility compared with the first set, it has to be asked, in light of these near-identical radiocarbon results, whether the first fluorine tests were not, after all, in this respect at least, "better" than the second?

Since 1959, the Piltdown material has twice been redated by the radiocarbon method.[73] The first of these tests, done in 1964, reportedly yielded exactly the same numbers as were obtained in 1959.[74] The second, and much more recent, test (and therefore, surely, that much more "improved") did not.[75] This 1989 test was done to assess whether the cranial fragments of Piltdown 2 could belong to Piltdown 1, as was strongly suggested by the 1955 report.[76] It involved testing a Piltdown 2 bone (for the first time) alongside the Piltdown 1 mandible, which was retested. (I do not know why the mandible rather than the cranium of Piltdown 1 was tested, given the stated rationale. But let that pass.) The results were as follows:

Piltdown 2 cranium   970 ± 140 years
Piltdown 1 mandible   90 ± 120 years.

Spencer notes that these results imply that the cranial fragments of Piltdown 1 and 2 "might well belong to two quite *distinct* individuals," after all; but what he does *not* comment on is the new, younger age (so young, indeed, that it may even be a pre-conception) of the forever-younger mandible.[77] Nor does he explain why these later (and therefore better) tests had a higher (and more variable) margin of error than the earlier ones.

And that's it. That's all I've got. "Of course, I've only scratched the surface" (but us postclassical theorists [theorist? where's the theory here?] will tell you that the surface, the appearance, is all).

So okay, I imagine you're asking yourselves, what do all these details add up to? Well, I warned you. "No denouement, no revelation." This text performs what it is about, that's all. Detail is detail.

NOTES

1   On the first project, see Malcolm Ashmore, "The Theatre of the Blind," *Social Studies of Science* 23 (1993): 67–106; see also Jan Sapp, *Where the Truth Lies: Franz Moewus and the Origins of Molecular Biology* (Cambridge, 1990). On the second, see Malcolm Ashmore, Michael Mulkay, and Trevor Pinch, *Health and Efficiency: A Sociology of Health Economics* (Milton Keynes, UK, 1989), chapter 7; see also Jonathan Potter, Margaret Wetherell, and Andrew Chitty, "Quantification Rhetoric — Cancer on Television," *Discourse and Society* 2 (1991): 333–65.
2   John Ziman, "Some Pathologies of the Scientific Life," *Nature* 227 (1970): 996; cited, inter alia, in William Broad and Nicholas Wade, *Betrayers of the Truth: Fraud and Deceit in Science* (Oxford, 1982), 83. For representative works on the first career

of the bones, see Arthur Keith, *The Antiquity of Man*, 2 vols., 2d ed. (London, 1929 [1915]); and Arthur Smith Woodward, *The Earliest Englishman* (London, 1948). On their second career, see Charles Blinderman, *The Piltdown Inquest* (Buffalo, 1986); Ronald Millar, *The Piltdown Men* (St. Albans, UK, 1972); Frank Spencer, *Piltdown: A Scientific Forgery* (London, Oxford, and New York, 1990); and Joseph Weiner, *The Piltdown Forgery*, 2d ed. (Chicago, 1980 [Oxford, 1955]).

3 E. A. Hooton, "Comments on the Piltdown Affair," *American Anthropologist* 56 (1954): 289.

4 Wilfred Le Gros Clark; quoted in Roger Lewin, *Bones of Contention* (New York, 1987), 75.

5 See Michael Mulkay and Nigel Gilbert, "Accounting for Error," *Sociology* 16 (1982): 165–83.

6 "The Piltdown Bones and 'Implements,'" *Nature*, 10 July 1954, 61.

7 A rather less laudatory account of these figures is given in some private correspondence of the time: "There was no Piltdown forgery. . . . With . . . [Oakley] as the mouth-piece . . . Le Gros Clark the windbag and Weiner as the garbage collector . . . I have got them 'holed' and am biding my time." See Alvan Marston to A. Gunner, 17 May 1955; quoted in Spencer, *Piltdown*, 229 n. 20.

8 Woodward, *Earliest Englishman*.

9 See David Waterston, "The Piltdown Mandible," *Nature* 92 (1913): 319; and Gerrit S. Miller, "The Jaw of the Piltdown Man," *Smithsonian Miscellaneous Collections* 65 (1915): 1–31.

10 See Kenneth Oakley and C. Randall Hoskins, "New Evidence on the Antiquity of Piltdown Man," *Nature*, 11 March 1950, 379–82. Pre–fluorine test estimates of the bones' age varied from 200,000 to 1,000,000 years. See Kenneth Oakley, "Dating Fossil Human Remains," in *Anthropology Today*, ed. A. L. Kroeber (Chicago, 1953), 47.

11 John Reader, *Missing Links: The Hunt for Earliest Man* (London, 1988), 76.

12 Ibid. See also William L. Straus, Jr., "The Great Piltdown Hoax," *Science*, 26 February 1954, 266.

13 See Spencer, *Piltdown*, 133.

14 Weiner, *Piltdown Forgery*, 30. See also G. Ainsworth Harrison, "J. S. Weiner and the Exposure of the Piltdown Forgery," *Antiquity* 57 (March 1983): 46–48.

15 Kenneth Oakley to Arthur Keith, 3 January 1950; quoted in Frank Spencer, *The Piltdown Papers 1908–1955* (London, Oxford, and New York, 1990), 189.

16 See Weiner, *Piltdown Forgery*, 26–35; and Joseph Weiner, Wilfred Le Gros Clark, Kenneth Oakley, G. F. Claringbull and M. H. Hey, F. H. Edmunds, S. H. U. Bowie and C. F. Davidson, C. F. M. Fryd, A. D. Baynes-Cope, and A. E. A. Werner and R. J. Plesters, "Further Contributions to the Solution of the Piltdown Problem," *Bulletin of the British Museum of Natural History (Geology)* 2, 6 (1955): 233.

17 See, for example, Kenneth Oakley and Joseph Weiner, "Piltdown Man," *American Scientist* 4 (October 1955): 577.

18 Spencer, *Piltdown*, 137. See also Harrison, "J. S. Weiner," 46–47.

19 Harrison, "J. S. Weiner" (quoting Weiner), 47.

20 Joseph Weiner, Kenneth Oakley, and Wilfred Le Gros Clark, "The Solution of the Piltdown Problem," *Bulletin of the British Museum of Natural History (Geology)* 2, 3 (1953): 141–46.

21 Weiner et al., "Further Contributions," 228.

22 See Alvan T. Marston, "Missing Link — But He Wasn't a Fake, Says Alvan T. Marston

FDS," *Picture Post*, 19 December 1953, 41–43; and Alvan T. Marston, "Comments on 'The Solution of the Piltdown Problem,' " *Proceedings of the Royal Society of Medicine (Section of Odontology)* 47 (1954): 100–102. I am preparing another paper on Piltdown which, among other things, deals in more detail with the arguments of this lone dissenter.

23   Sherwood Washburn, "An Old Theory Is Supported by New Evidence and New Methods," *American Anthropologist* 56 (1954): 437; his emphases.

24   A. Carnot, "Recherches sur la composition générale et la teneur en fluor des os modernes et des os fossiles de différents âges," *Annales des Mines (Paris)* 3 (1893): 155–95; and J. Middleton, "On Fluorine in Bones, Its Source, and Its Application to the Determination of the Geological Age of Fossil Bones," *Proceedings of the Geological Society of London* 4 (1844): 431–33. For some of the accounting strategies associated with "neglect" and "rediscovery," see Augustine Brannigan, *The Social Basis of Scientific Discoveries* (Cambridge, 1981), chapter 6. For accounts of fluorine dating and its Piltdown connections, see Reader, *Missing Links*, 73–77; Blinderman, *Piltdown Inquest*, 65–69; Spencer, *Piltdown*, 128–29; and Sonia Cole, *Counterfeit* (London, 1955), 152–56.

25   According to Spencer, in *Piltdown*, Oakley's first use of the test, on the contentious Kanam-Kanjera remains discovered in Kenya by Leakey, was *not* successful "because of excessive amounts of background fluorine in the . . . material" (129). On the Galley Hill skeleton, see Kenneth Oakley and M. F. A. Montague, "A Re-Consideration of the Galley Hill Skeleton," *Bulletin of the British Museum of Natural History (Geology)* 1, 2 (1949): 25–46. The uncertainty and circularity of whether an experiment tests a phenomenon (or theory) or tests "the validity of [a] methodology" (Spencer, *Piltdown*, 129) or the competence of an experimenter or theorist has been stressed by Harry Collins, *Changing Order: Replication and Induction in Scientific Practice* (London and Beverly Hills, 1985); and by Trevor Pinch, "Theory Testing in Science — the Case of Solar Neutrinos: Do Crucial Experiments Test Theories or Theorists?" *Philosophy of the Social Sciences* 15 (1985): 167–87.

26   Millar, *Piltdown Men*, 194; my emphases.

27   See Alvan T. Marston, "The Relative Ages of the Swanscombe and Piltdown Skulls, with Special Reference to the Results of the Fluorine Estimation Tests," *British Dental Journal* 88 (1950): 299. For the Galley Hill dating, see Oakley and Montague, "A Re-Consideration."

28   Oakley, "Dating Fossil Human Remains," 46, 47.

29   H. de Vries and Kenneth Oakley, "Radiocarbon Dating of the Piltdown Skull and Jaw," *Nature*, 25 July 1959, 225. It could be argued that it was only *before* 1953 that Oakley can be said to have *known* the source of the Piltdown bones: namely, the Piltdown gravels. Once the forgery hypothesis was in place, their source became radically unknown. The dating carried out in 1953 thus cannot, by this argument, be taken to be "precise."

30   Robert F. Heizer and Sherburne F. Cook, "Comments on the Piltdown Remains," *American Anthropologist* 56 (1954): 93.

31   Cole, in *Counterfeit*, hints at a third claim for improvement: the use, in 1953, of a "blind" analysis. The samples "were contained in little glass tubes, each with a number. The chemists who did the tests had little idea what the samples were or where they came from" (156). Although she does not say it in so many words, the implication is that the 1949 test was not carried out in this rigorous fashion.

32   Weiner, Oakley, and Clark, "The Solution," 143.

33   Weiner et al., "Further Contributions," 256.

34   For an analysis of the rhetorical utility of other forms of "undeniability device," see Derek Edwards, Malcolm Ashmore, and Jonathan Potter, "Death and Furniture: On the Rhetoric, Politics and Theology of Bottom-Line Arguments Against Relativism," *History of the Human Sciences* 7 (in press).

35   Oakley and Hoskins, "New Evidence," 381. For some reason, this figure was given in the text, but not in the table.

36   This rhetorical effect is spoiled somewhat by the use of considerably more "accurate" figures in a 1935 text on the measurement of fluorine content in (modern) human teeth (never otherwise cited in this literature); see Marston, "Comments," 101, citing P. J. Brekhus and W. D. Armstrong, "A Method for the Separation of Enamel, Dentine and Cementum," *Journal of Dental Research* 15 (1935): 23. (Their figures go to *four* decimal places — very impressive for such an "early" work!)

37   See Weiner et al., "Further Contributions," 256.

38   Cole, *Counterfeit*, 154, 153.

39   See Weiner, *Piltdown Forgery*, 30.

40   Spencer, *Piltdown*, 137.

41   Weiner et al., "Further Contributions," 256; my emphases.

42   A similar form of circularity, known as "the experimenters' regress," which is endemic to the experimental determination of novel phenomena, is beautifully explored in Collins, *Changing Order*, chapters 4 and 5.

43   The single exception, as far as I know, appears in the 1959 account of the first radiocarbon dating; see de Vries and Oakley, "Radiocarbon Dating," 224. Here, the two 1950 values (0.1% and 0.2%) were ascribed, re(tro)spectively, to the cranium and the mandible. (A possible rhetorical explanation for this is explored below.)

44   Oakley and Hoskins, "New Evidence," 380; my emphases.

45   Weiner, Oakley, and Clark, "The Solution," 143.

46   Oakley and Weiner, "Piltdown Man," 577. Note that "amounts" is ambiguous here: it could refer either to sampled material or to fluorine. For the purposes of my argument, I am treating it as the former.

47   See de Vries and Oakley, "Radiocarbon Dating," 224.

48   See Joseph Weiner and Kenneth Oakley, "The Piltdown Fraud: Available Evidence Reviewed," *American Journal of Physical Anthropology* 12 (1954): 2.

49   In 1953, Oakley stated that seventeen samples of *Eoanthropus* had been analyzed; see his "Dating Fossil Human Remains," 52. Cf. Table 1, col. 1, and note *e*.

50   Sherwood Washburn, "The Piltdown Hoax," *American Anthropologist* 55 (1953): 760.

51   Interestingly, this bone has been interpreted as a fifth piece of the same skull used for the Piltdown 1 assemblage; see Oakley and Weiner, "Piltdown Man," 583. Recent radiometric dating results have cast doubt on this interpretation; see Spencer, *Piltdown*, 230 n. 30.

52   See Oakley and Montague, "A Re-Consideration"; and Kenneth Oakley, C. Randall Hoskins, and G. Henri-Martin, "Application du test de la fluorine aux crânes de Fontéchavade," *L'Anthropologie* 55 (1951): 239–42. See also Oakley, "Dating Fossil Human Remains," 45–46; and Robert W. Ehrich and Gerald M. Henderson, "Concerning the Piltdown Hoax and the Rise of a New Dogmatism," *American Anthropologist* 56 (1954): 433–35. Note that Ehrich and Henderson do not themselves question the validity of these datings; indeed, they describe them as "soundly dated" (435).

53 Weiner, Oakley, and Clark, "The Solution," 143.

54 See Oakley and Weiner, "Piltdown Man," 577. Elsewhere, an amusing account of how and why these supposedly larger samples were now available was offered: "Since Piltdown man had shed his aura of extreme antiquity after [1949], drilling could this time be carried out more boldly"; see Cole, *Counterfeit*, 155.

55 De Vries and Oakley, "Radiocarbon Dating," 224.

56 Weiner and Oakley, "Piltdown Fraud," 2.

57 C. F. M. Fryd, "Chemical Changes in Bones: A Note on the Analyses," in Weiner et al., "Further Contributions," 266.

58 Ibid.

59 C. R. Hoskins and C. F. M. Fryd, "The Determination of Fluorine in Piltdown and Related Fossils," *Journal of Applied Chemistry* 5 (1955): 86–87.

60 Oakley, "Dating Fossil Human Remains," 52.

61 Oakley and Hoskins, "New Evidence," 380; my emphasis.

62 See Weiner, Oakley, and Clark, "The Solution," 143; and Weiner et al., "Further Contributions," 260, 262–65.

63 Specifically, there is no mention of fluorine/phosphate ratios in Blinderman, *Piltdown Inquest*; Cole, *Counterfeit*; Reader, *Missing Links*; or Spencer, *Piltdown*, all of which discuss fluorine-content dating at length.

64 Spencer, *Piltdown*, 139.

65 See Heizer and Cook, "Comments," 93.

66 Washburn, "Piltdown Hoax."

67 Heizer and Cook, "Comments," 93.

68 Weiner and Oakley, "Piltdown Fraud," 4.

69 Oakley and Weiner, "Piltdown Man," 582; my emphasis.

70 De Vries and Oakley, "Radiocarbon Dating," 224.

71 Ibid. See also W. M. Krogman, "The Planned Planting of Piltdown: Who? Why?" in *Human Evolution: Biosocial Perspectives*, ed. S. L. Washburn and E. R. McCown (Menlo Park, CA, 1978), 238–51.

72 Weiner et al., "Further Contributions," 257.

73 See Spencer, *Piltdown*, 229–30 n. 30.

74 See J. C. Vogel and H. T. Waterbolk, "Groningen Radiocarbon Dates [Piltdown Series]," *Radiocarbon* 6 (1964): 368. See also Spencer, *Piltdown Papers*, 198 n. 6.

75 See R. E. M. Hedges, R. A. Housley, I. A. Law, and C. R. Bronk, "Radiocarbon Dates from the Oxford AMS System: Archaeometry Datelist 9," *Archaeometry* 31 (1989): 207–34 (contribution of Frank Spencer and C. Stringer).

76 Weiner et al., "Further Contributions," 260.

77 Spencer, *Piltdown*, 230 n. 30.

A Glance at SunSet: Numerical Fundaments in

Frege, Wittgenstein, Shakespeare, Beckett

John Vignaux Smyth

For what are things independent of the reason? To answer that would be as much as to judge without judging, or to wash the fur without wetting it. — Gottlob Frege

In the domain of contingency, the *point of origin is a missing point*. . . . The necessary universal reference is at infinity, beyond history and delivered from space. Contrary to Pascal, Leibniz leans on the calculus and on geometry to conserve the continuity between the two worlds, and the "true point of view of things" is of *the same ordonnance,* as we have seen, as the starry site [*site etoilé*] of our errors, our evils and our pains. — Michel Serres

For Two is the most imperfect of all numbers. — Christopher Smart

And when it comes to neglecting fundamentals, I think I have nothing to learn, and indeed I confuse them with accidentals. — Samuel Beckett

ARISTOTLE somewhere compares the middle term of a syllogism both to the moon, between earth and sun, and to the situation of two friends who share an enemy in common. The seemingly haphazard connection made by these two analogies would appeal to René Girard, whose anthropology is founded on the twin principles of *mimesis* and *exclusion:* in brief, the "mimetic scapegoating" of an arbitrarily selected enemy or victim whose exclusion from the order of the living ("friends," tribal order) provides the foundation or *fundament* of that order, its false sense of transcendence, and the source of universal religious sacrifice.[1]

I should explain that the first pun in my title, "SunSet," exploits Frege's notorious abandonment of the grand project of reducing number to set theory. But my pun also alludes to the fact that the moon — mythic emblem of reflection and mimesis, and especially of the ambivalence surrounding relations of identity/difference — rises prominently, almost obsessively, in works by all of the authors to whom I refer in this paper: Frege, Husserl, Wittgenstein, Shakespeare, and Beckett.[2] My second pun, "fundament," is adapted from my Beckett epigraph.

I read Beckett's joke in *Molloy,* that the moon always shows us her arse, and Shakespeare's joke, in his play with Bottom, that the moon has bored a hole through the earth to take the place of the sun, in the same way. The moon that rises at SunSet is in both cases a *mimetic fundament*—rather as Wittgenstein would also speak, once he had rejected Russell and Frege, of mimesis and exclusion in terms of the "Bottom" of mathematical calculation. I am not an orthodox Wittgensteinian of any stripe; I think he has been justly criticized (as well as praised) for being subject to the "aberrances of the poets."[3] But I do want to show, precisely, that there is an important convergence of literature and the philosophy of mathematics when it comes to "fundamental reflections."

Although I shall barely deal with Girard directly, my attempt here to navigate the (obviously uncircumnavigable) topics of mimesis, order, and exclusion in both literature and the philosophy of mathematics is to some extent inspired by Girardian mimetic models (of which more in a moment). In particular, I view the fundamental relation of identity/difference as mediated by similarity or likeness, so that order or differentiation is a triple relation made up of *two* differences: one between the same and the similar, and another between the similar and the dissimilar. Order (so to speak, with a capital *O*) is thus composed of the domain of *more or less* (order with a lower-case *o*), such as the more or less lovable, as differentiated from the domain of exclusion, such as the *not*-lovable. The relation of exclusion is more primitive than lower-case ordering in the sense that the included or excluded elements are not necessarily ordered among themselves or in relation to each other. The relation(s) between lower-case order and exclusion is illustrated, for example, by the relation(s) between the relations among the integers and the relations between numbers and not-numbers.[4]

Another basic perspective here, consistent with Girard but trans-Girardian, is a picture of the road to truth as proceeding by way of the exclusion of falsehood, but where the true tends to get excluded along with the false and the history of the exclusion of falsehood is by no means linear. Excluded middles and apagogic proofs (by reductio ad absurdum), of course, illustrate the principle of logical exclusion in a standard "exhaustive" way (hygienically excluding the false with no remainder). But Beckett's Molloy speaks more to my immediate purpose when he claims that "all that is false may more readily be reduced, to notions clear and distinct," than what is true. "I think so, yes. . . . But I may be wrong."[5] (The madman who claims to be able to iterate π by heart can never prove himself right, although we may prove him wrong.)

More narrowly, in Girard's anthropology, the false transcendence of the excluded scapegoat-divinity is never *entirely* false: source of sacred ambivalence, in Durkheim's sense, of good and evil, order and disorder, and so on, the foundational principle is interpreted by primitive man (by "sacrificial" man more generally) in terms which are *both true and false* (and which, in Girard's hypothesis, are basic to the original development of conceptions of true and false). Girard aside, however — and although I do not deny the kind of asymmetry between truth and falsity which sustains, say, the celebrated scientific Principle of Falsifiability or Molloy's adage about falsity's reducibility to "distinctness" — the Leibnizian and Beckettian views cited in my epigraphs suggest my general perspective (as well as adumbrating Molloy's ironic caveat), where true and false, alongside contingent and universal, accidental and fundamental, are said to be of "the same ordonnance," inextricably mixed.[6] This necessary mixture is most obvious in social relations (since the social order is transparently founded on lies and concealments as well as truth-tellings), but it is surely not exclusive to them.[7] The shape of the mixture might indeed be said to define the shared limit, the momentary equilibrium or "saddle point,"[8] of the shifting interaction between social or "intersubjective" constructions and the progress (or regress) of "objective" claims.

Although the work of Michel Serres on the history of science and mathematics is not presupposed here, his "sacrificial" account, in *Les Origines de la géométrie*, of Western mathematics is worth mentioning. Beginning with a Girardian analysis of the relation of mimesis, the sun, and burial grounds in the work of Thales (discoverer of triangular similarity), Serres goes on to describe the first known apagogic proof in history, that of the irrationality of root 2, in explicitly Girardian terms. Here, the "objective" discovery of proof *by exclusion* coincides with social, verbal, and cognitive exclusions of a less hygienic sort. Serres not only sees in a Girardian light the mythic (and perhaps also real) deaths associated with the Pythagorean discovery of mathematical irrationality, and the secrecy-on-pain-of-death surrounding it, but he also exploits the mimetic dimension of the difference between even and odd numbers on which the classical proof depends. The association of "even" with sameness, rule or model, as opposed to the "odd" exception, is, of course, built into English etymology and even grammar. Its association with mimetic resemblance is certainly to be found in literature — by Flann O'Brien, Nabokov, Beckett, Shakespeare, and many others — and perhaps also in the proverbial evil of evens, evenings and twilights, as well as the primitive terror of twins. For Girard this terror is the mythic projection of a primordial insight into the threat posed by "mimetic desire," where imitation between two or more parties is liable to generate violent rivalry over their mutually desired objects. (The mass crises provoked by such rivalry, in Girard's hypothesis, are resolved by the mimetic polarization of violence against a single victim or small group, the first "sacrifice.") But the rub of Serres's Girardian argument (backed up by an acute analysis of Socrates' discussion of geometry in the *Meno*) is that the general suspicion and "sacrifice" of number provoked by the first apagogic proof provided the philosophic *foundations* of Euclid's determined attempt at "purely" geometrical rigor — the paragon of rigor that Frege, in *The Foundations of Arithmetic,* wished to be the first to match in the domain of number. Serres's Girardian moral: truth and its sacrifice march hand in hand.[9]

So much for Serres's picture of mathematical history. Now let's consider two pictures that illustrate my own kind of appeal to "mimetic models."

The first picture or puzzle worried Charles Dodgson: a circular puzzle which will serve us well again when we come to Wittgenstein, Shakespeare, and Beckett. If one leaves London at midday, traveling west "at the speed of the sun," so that it remains forever midday, at what point does one change the date on one's watch? Clearly, conventional world-time is arbitrary in its ordering. Further, if at any given moment local times are globally assigned, we are seemingly forced to assign a double and thus "contradictory" time-date to some point (say, at midnight in the Pacific): not merely the definitional $(24\text{hrs}, X\text{th}) = (0, X + 1)$, but $(0, X) = (0, X + 1)$. However, by arbitrarily assigning one of the two possible time-dates and excluding the other, we can make the contradiction appear to disappear. But in this case we are forced to deny, in effect, either that one day has a definable beginning or that the other has a definable end.

Postcolonially speaking, Greenwich Mean Time is, of course, not so geographically "arbitrary" as all that! I simply wish to illustrate the consequences here of what Girard calls mutual *negative mimesis*, the agreement to be different (to posit different times), that is, to illustrate how a negative mimetic model of the kind also important to Wittgenstein, Shakespeare, and Beckett may generate foundational points, or *fundaments*, of exclusion.

My second picture is no looking-glass puzzle, but one borrowed from the recent, rather epistemologically optimistic *Reality of Numbers* by John Bigelow. It is the simple "instantiation," as performed by Bigelow, of the imaginary numbers in terms of an idealized kinship structure (following the lead of Quine's instantiation of the rational numbers in terms of ancestor relations), where kinship relations are made to correspond by definition to relations in the imaginary field (usually modeled, of course, in two-dimensional geometry). All that is required is that, when Bigelow's model tribespeople enumerate all their kin, we suppose that they do so in a conventionally defined order and that, "to simplify matters, we suppose that intermarriage never occurs across generations. Neither Adam nor Eve coupled with their children," and so on.[10] Imaginary kinship here depends simply on a conventional order and a rule of exclusion: a most ideal Girardian combination, quite literally, of mimesis and incest prohibition. For Bigelow, moreover, this "instantiation" of the imaginaries demonstrates that they are more than merely phantoms given artificial life in Cartesian geometry, but rather belong to the "reality" and "universality" of number understood in terms of *relation* more generally. Bigelow's ideal bridging of social and mathematical worlds in terms of mimesis and exclusion may prepare us imaginatively for the more "skeptical" bridging attempted by Wittgenstein.

2. But it makes little sense to discuss Wittgenstein's at least double (early and late) philosophies of mathematics without at least alluding to early Frege and Russell—who were excluded, in the sense of being named as opponents, by the later Wittgenstein of the *Remarks on the Foundations of Mathematics*. For symmetry, I also allude to the early Husserl (who was excluded, in the sense of being ignored, in the *Remarks*). As a matter of fact, not only do Frege's and Husserl's respectively logical and psychological attempts to "found" or "ground" arithmetic appear to exclude each other, but both result in the relation of exclusion itself being placed center stage. This is no accident, perhaps, inasmuch as the battle between "logic" and "psychology" can itself be perceived in terms of the classic psycho-logic of negative mimesis, that is, self-definition by means of exclusion of the enemy.

In his early "Concept of Number," Husserl claimed that the a priori foundation of all judgments of numerical identity/difference must be an act of pure exclusion ("psychical segregation") which divides a set into (at least) two subsets "prior" to any judgments of similarity/difference. Logical judgments are made merely "for the ulterior purposes of thought, to segregate similar things, . . . absolutely not to separate *for the first time* what originally is" an identical unity (an act that is purely "external," the task of segregation-exclusion proper).[11] Ordinary judgments of similarity/difference and "segregations," although conflated "as a rule," in principle entail "two wholly different kinds of inclusion." In principle, segregation/exclusion accounts for the void, zero, and is itself void of "positive" (here also called "physical") content, an act of apparently pure negation — or pure "form." It is an apparently incidental but explicit result of this view that arbitrary acts of "psychical segregation" are *always justified,* in that *some* genuine distinction among their contents can always be maintained retrospectively. Arbitrariness thus always "contains" necessity and truth (number itself) — although Husserl made no mention of the inverse proposition nor, in this essay, of falsity.[12] We shall see in Beckett's work how both falsity and arbitrariness play a determining role in what Husserl suggestively called the "ulterior purposes" of our differentiation of similars. The "ulterior" (of Husserlian difference) and the "exterior" (of Husserlian segregation), I suggest, are potentially too complicitous for Husserlian comfort.

Frege would more than agree with me, for in his stylish *Foundations of Arithmetic* he dismissed all psychological foundations of number as begging the fundamental logical question, viewed in terms of resolving the identity crisis besetting all explanations of numerical iteration prior to his own. But precisely where the psychological or "subjective" problem was excluded, a catastrophically "objective" version of the exclusion problem appeared, by poetic justice, namely, the notorious Russell paradox (concerning sets *that exclude themselves*), which led Frege to abandon the project of the *Foundations.* As is well known, Frege rejected the iterative "indexing" solution to the paradox, known as the theory of types — which the later Wittgenstein also rather astonishingly called "a lie" — preferring to sacrifice his whole programme (and turning at last to putative geometrical foundations of arithmetic).

This rejection was the consequence of his equally well-known refusal to sacrifice the concepts of *concept* and *object* as distinct from set theory. At the same time, the obvious analogy between concept and set — the concept "sheep" seems to have as its extension the "set of all (possible) sheep" — suggests why the theory of types (indexing metalevels to exclude mutual contamination) was anathema to him, since the opening of the *Foundations* famously proclaims: "It is a mere illusion to suppose that a concept can be made an object without altering it." (Wittgenstein later agreed that the concept "concept" was "vague.")[13] Frege hoped that this exclusion of metalevel iteration would hold generally for concepts but *not* for sets; and Russell's paradox could no more be solved for him by indexing sets than his own concept-object (or metalevel concept) prohibition could be circumvented by indexing metalevel concepts. For Frege, using the theory of types to ground the number system was merely to go in a vicious circle, hoping to explain one form of iteration by another. Moreover, the theory of types is founded on *two* orders: that of the binary type distinction-*exclusion* (of particulars and classes), supplemented by an iterative ordering of classes.[14] (These distinctions follow the exclusive and inclusive orderings posited in my opening discussion.) One should also note that avoidance of the paradox in von Neumann-style set theory entails the *philosophically* dramatic exclusion of anything but sets from the system, that is, the exclusion of elements of sets or particulars *tout court*. In terms of the concept/set analogy, this would mean a world of concepts without objects.

Frege's rejection of such formalist solutions stemmed from his refusal to regard philosophy (the world of concepts) and mathematical logic (the world of sets) as mutually exclusive. It is more than merely curious, then, that the Russellian trouble with metalevel sets finds an important, though easily muffled echo in Frege's own treatment of concepts. I have no space here to analyze the complex interplay of logical and rhetorical moves in the extraordinary passage (whose springboard is the ontological argument for the existence of God), which begins with the claim that "existence is in fact nothing but denial of the number nought," and which generates Frege's only treatment, in the *Foundations*, of a "so to say, second order concept," where, for example, "oneness is a component characteristic" of the "concept [of] all concepts [e.g., "moon of Earth"] under which there falls only one object."[15] "Under certain conditions," we are told—one is tempted to say *contingent on the a priori situation* or *once in a blue moon*—the internal "component characteristics" of concepts entail their external properties (which depend, like the properties of existence, on objects), and vice versa, "just as we can *occasionally* infer the durability [or desirability] of a building from the type of stone used in building it."[16] However, later we also read that analytic conclusions ("which extend our knowledge, and ought therefore, on Kant's view, to be regarded as synthetic") "are contained in the definitions, but as plants are contained in their seeds, not as beams are contained in houses."[17]

Without claiming to demonstrate them, I merely suggest here that such complex shifts between concept and object, internal and external, analytic and synthetic, and so on, generate, not accidentally, at least some trace of an inversion of the initial rhetorical alignment of the "component characteristics" of concepts (such as definitions) with beams rather than houses. But it would take a lengthy Derridean-style reading to make a plausible case for the specter of beams or walls *both inside and outside* their houses—prophetic of Russell's sets that both include and exclude themselves — as no more of a merely rhetorical accident than Frege's conventional shift to an organic analogy which seems to lay that specter to rest.[18]

Fortunately, Frege provided a briefer way to summarize the problem of exclusion in terms of the relation between concepts and numbers without forcing us into the arcana of set theory, on the one hand, or the minutiae of deconstructivist analysis, on the other. Frege's refusal to exclude zero (or one) from equality of status with other numbers — let alone, to grant it metalevel status[19] — followed from his definition of zero, which, for analytic convenience, he defined as the number belonging to the concept "not identical with itself." But since a number x is said to "belong" to a concept by virtue of that concept's having x objects, *any* concept without object would serve. Thus zero buys its uniqueness as object *and* its "equality" with every other natural number at the price (or offering the gratuitous gift) of the total undifferentiation not only of the logical and empirical, but also of false and incoherent concepts. No concept to which zero "belongs" can be said to be viciously incoherent, any more than Husserlian segregation can ever be false. (The line carefully established earlier between a blue moon and the "nonsense" of a blue concept — Frege's example of an a priori error — is in this context erased.) *Nothing is properly excluded from reason,* as Frege suggested in my epigraph, *once Nothing has been properly included.*[20]

This may sound too much like a Beckettian or Shakespearean pun to convince logicians or philosophers of its significance. Yet Frege's jealous patrolling of the border which excludes psychology, anthropology, and literature from the realm of logic and philosophy undeniably generated and safeguarded a most *inclusive* and rich view of logic and philosophy "themselves." Thus Frege's scrupulously logical scheme explicitly allows for even more radical kinds of undifferentiation than that suggested by poetic Zeros — or even by Unique Blue Moons rising over Beams in Desirable Houses! We are soberly told, for instance, that nothing in logical principle prevents numbers from being called "wider" (or narrower) than nonnumerical extensions of concepts.[21] And that "if n is not a number, then n itself is the only member of the series of natural numbers ending with n, — if that is not too shocking a way of putting it."[22] Far from confirming the common contemporary view of Frege's parochial logicism, such apparently "nonsensical" remarks (readers may check the text for their logical propriety) represent a most shockingly inclusive idea of number, one that is potentially radical in terms of both ontology and epistemology. These remarks may also demonstrate to what degree

Frege's heroic attempts, contra Husserl, *to exclude arbitrary exclusions* seem prescient of the cognitive threat that would shortly materialize for him in the Russell paradox: the threat of exclusive relations as such, perhaps, to *any* project that would claim theoretical unity.

3. Although the relation between concept and object is classically defined, in both empiricist and rationalist traditions, in terms of mimesis, we have not yet actually encountered mimesis, in either the Girardian sense (imitation of behavior) or the classical sense of representation.[23] But in Wittgenstein, however otherwise un-Girardian, the dramatic conjunction of these two senses of mimesis is unavoidable. The relation between them is already a drastic problem in the *Tractatus,* with its opposition between the "picture-theory" of the speakable and the famous philosophic prohibition of the unspeakable but "shewable."[24] Here is an explicit crisis of the relation between mimesis-as-representation and mimesis-as-imitation — one which is supposed to provide the foundations of the *end* or *bottom* of philosophy to boot!

But here I am primarily concerned with three central topics of the *Remarks on the Foundations of Mathematics* — agreement, rule, and miscalculation — where allusions to "copyings" of every kind are ubiquitous, even obsessive. No less so is the insistence on conceiving of order as founded on exclusions, exclusions which may obscure the extent to which "our mathematics is built up on . . . an *unordered* generality."[25] Where the *Tractatus* speaks of ascending the ladder of the propositions to the unspeakable (before throwing it away), the *Remarks* suggest that, far from excluding Russell's paradox of exclusion, we might *begin* from it, as "something that towers above the propositions and looks in both directions like a Janus head. . . . And as it were descend from it to the propositions."[26] Similarly, Wittgenstein notoriously suggested that we reverse the standard view of the number system, whereby imaginary includes real includes rational, and so on. If anything, he maintained something like the historical perspective in which each new domain is always viewed from the foundational standpoint of what was hitherto prohibited and excluded (e.g., seeking the square root of 2 or −1), now converted like the Girardian scapegoat into the source of a new but similarly false or potentially arbitrary transcendence: the "objective" vision of number as such.[27] He would thus obsessively return, at the limit, to the possibility of an arithmetic which could include division by zero: an arithmetic founded on *the* fundamental arithmetical exclusion. In such a context, a Girardian like Serres could not fail to be struck by Wittgenstein's repeated suggestion that we might "calculate as we do calculate (all agreeing), etc., and yet at every step have the feeling of being guided by the rule as if by a spell. . . . (Thanking the deity perhaps for this agreement.)"[28] The "peace" which Wittgenstein and Serres stress as surrounding mathematical certainty, our "magical" agreement, is analogous to the peace which Girard claims to derive from the exclusion of the arbitrary scapegoat-divinity, the "excluded third." Logicians, according to Wittgenstein, confuse the essential with the accidental;[29] and he consistently conceived of the attempt to exclude falsehood or contradiction as a misguided attempt to exclude arbitrariness or contingency. More strikingly still, he compared division by zero to the situation of tribespeople who, failing to pick up dropped coins, say, " 'It belongs to the others' or the like."[30] Reciprocity between true and false, law and arbitrariness, thus explicitly becomes a contingent (but potentially global) reciprocity between self and other.

If agreement were the sole foundation of identity and truth, then the wholly "conventionalist" reading of Wittgenstein would go unchallenged, and truth might be indistinguishable from imitation pure and simple. But this is vehemently denied by Wittgenstein in his assertion that "nothing" can be regarded as the foundation of our criteria of identity, even though mathematical order is explicitly compared to "fashion."[31] I am reminded of Shakespeare's *Much Ado About Nothing*, where fashion (said to be "nothing to a man" — the Law of "Asses," or "Bottoms") is explicitly founded on the arbitrary exclusion of "one Deformed." Nevertheless, like Shakespeare and Beckett, Wittgenstein refused to make the pseudo-Nietzschean move of reducing everything to "appearance," but rather insisted on the coincidence of appearance and reality: "Mathematical propositions seem to treat neither of signs nor of human beings, and therefore they *do* not." Yet this apparently "objectivist" point is made elliptically, to say the least, complicated once again by a psycho-logic of exclusion: "We acknowledge it [the point] *by* turning our back on it."[32]

The argument is curious. Although the "cousins" of agreement and rule are the twin foundations of the *Remarks*, it is following a rule, not agreement, that is said to be "FUNDAMENTAL," or "the *bottom*" of our language games.[33] For while "admittedly going according to a rule is also founded on agreement, . . . one does not learn to obey a rule by first learning the use of the word 'agreement,' " but vice versa, and "to understand what it means 'to follow a rule,' you have already to be able to follow a rule." Similarly, "it could be said: that science would not function if we did not agree regarding the idea of agreement. . . . And it goes on like that."[34]

Thus agreement is *excluded* from laying claim to the title of "funda-ment" or "bottom" inasmuch as it always *includes* or presupposes itself circularly, or in an iteration without end. And since it seems that we are also forced to regard the concept of rule (like the concept "con-cept") as quite as "vague" as this endless iteration, the bottom of rule can only be automatic concord, or, as it is put in the context of the deity mentioned above, repetition of "the same," mechanically — not without thinking, but without *reflecting*. Indeed, Wittgenstein did assert that no more could be said about the ultimate meaning of rules than how they had been learned, defined primitively in terms "of shewing how and of imitation, of lucky and misfiring attempts, of reward and punishment and the like." [35] Mimesis-reflection gives way, at bottom, to mimesis-imitation.

But if positive imitation is thus crucial, it is not exhaustive. In accor-dance with his rather obsessively repeated exclusion of rules and games illustrated by only a single instance, Wittgenstein excluded from rule-learning the reproduction by one chimpanzee of a pattern of marks made *once* by another chimp (though the example of successive *mutual* mimeses between the chimps would seem to render this point some-what moot). Nevertheless, the main point is Wittgenstein's insistence not only on a process of mutual interaction or instruction, but also on the importance of the exclusion — which I will regard in terms of negative mimesis — of mistaken acts, of miscalculation, which "is really the key to an understanding of the 'foundations' of mathematics." This, then, is the *fundament* of "FUNDAMENTAL" rule, and it explains why, when push came to shove, Wittgenstein put such emphasis on punishment instead of reward, on the "peculiar attitude [learned by chil-dren] towards a mistake," on obeying out of fear, and even (fancifully) on ceasing to divide by zero because it had led to "the loss of human life." [36] (Such linkage between the exclusion of falsehood or arbitrariness and of pain might easily provoke a reader of Beckett to pay rather special atten-tion to the extraordinary passage in which Wittgenstein denies that it is mere "moonshine" to assert that calculation differs from "experiment," just as poetry is not "a psychological experiment" and pain "is not a form of behaviour." [37] I shall return to this later, for now observing only that calculation is described as "overdetermined," as standing "on four feet, not on three," [38] as circular in that the correctness of the calculation is judged by its result — in short, as *a self-fulfilling prophecy*.)

Negative mimesis (what not to imitate) is fundamental, but it does not by itself determine rule, the order to be imitated: thus the assertion that rules are "arbitrary," and thus Wittgenstein's dogged, almost insanely reiterative playing out of his "skeptical problem," which Saul Kripke identifies in terms of Nelson Goodman's paradox, that is, not only the impossibility of specifying a unique rule for any finite sequence, but the impossibility of ruling out wholly arbitrary hypotheses. *Every* rule is thus conceived of as a mere negation-exclusion, and this effectively amounts to implying that *any* extension of a given pattern may constitute metalevel presumption. Immediately before taking up Frege's claim about the inconceivability of non-self-identical objects, Wittgenstein asks: "Is something that follows from a rule itself in turn a rule? And if not, — what kind of proposition am I to call it?"[39] The prohibition of metalevel presumption that makes negative mimesis fundamental thus provokes an iterative identity crisis of the acute kind that Frege saw in work on number theory prior to his own, so that iterating the odd numbers is analagous to promising every day, "Tomorrow I will give up smoking": Is one always doing the same thing or something different?[40]

Moreover, a suspiciously similar, though much grander, mimetic crisis occurs even more violently in the heart of philosophy itself. Wittgenstein's celebrated philosophic prohibition of metalanguage, "preparatory language," and so on, surely amounted to no more or less than the sacrifice of all philosophical theory (what he called "explanation") on the very altar of classical mimesis-as-representation ("pure description"). This sacrifice was supposed to differentiate philosophy from science and logic, not, as in the *Tractatus,* by committing suicide and resorting to pure "shewing" (mimetic in the sense of imitable),[41] but nevertheless by pretending to its own variety of mimetic ("descriptive") centrality: a slow "cure" that would, in Wittgenstein's celebrated metaphor, purge the "disease" of philosophy itself. In the *Tractatus* a hygienic break had been made with (bad, spoken) philosophy in favor of the equally hygienic opposition between science and the unspeakable (all this hygiene being guaranteed by the sacrifice not only of philosophy, but of the famous Wittgensteinian ladder which has brought us to such heights). But in the later Wittgenstein, philosophy became frankly what both Girard and Derrida, following Plato's description of mimetic poetry in the *Republic,* call a *pharmakon:* a mimetic drug which is both remedy and poison. Like poetry in the *Republic,* philosophy must first be expelled as a *pharmakos* or scapegoat, then return ambivalently as a *pharmakon.* Since the "descriptive" philosophic cure would no longer be needed with the end of the "explanatory" philosophic disease, Wittgenstein could defer into the future the second sacrifice of philosophy, which mirrored the first in his own past. Nevertheless, in both cases it was the specter of this ultimate sacrifice which sustained philosophy as he so seductively and *doubly* practiced it!

These patterns are too compelling to ignore, providing as they do a deep interconnection among Wittgenstein's various so-called objectivist, intersubjectivist, and subjectivist tendencies. My opening illustrations of negative mimesis (Dodgson's circular time and Bigelow's fictive tribe) now pay off in assessing Wittgenstein's own relentless recourse to both circular and anthropological imagery in the *Remarks*. Not only does a simple picture of a circle of copied 2s emerge, apparently out of the blue (where "the last figure is a copy of the first one"[42]) and without any further explanation, immediately prior to yet another passage concerning the relation between calculation and prediction, but *graphic* circular imagery (as though his sketch might magically help us understand!) dominates the all-important topics of negation and contradiction. Thus Russell's paradox is said to be as "natural" as saying that a point lies left and right along a circle, while nonclassical double negation (remaining negative) is "represented," or rather literally drawn, as a repeated movement along the same half of a circle.[43] Immediately after this depiction of the Russell paradox, moreover, the thought-experiment of a logic of pure exclusion appears, where, by blue, say, I mean not-blue, since "the colour that appears to me always counts as the one that is *excluded*" (as though "God always shews me a colour in order to say: *not* this").[44] If this nicely suggests the Girardian principle of divinity — including the pharmakon-ambivalence by which "we could be led to want to describe something's being blue, both by saying it was blue, and by saying it was not blue"[45] — it is by no means singular in its anthropological suggestiveness. I note especially the example of the tribe in which the function of the chief is not necessarily to give orders, that is, where the principle of order *is excluded from ordering.*[46]

Space constraints preclude my treating Wittgenstein's examples and prodigal "imageries" here as just as fundamental as the arguments that appear to link them, thus applying to him his own wise advice to logicians that they reverse "essentials" and "accidentals." Such a project would proceed elsewhere from canonical interpretations, whether "objectivist" or "culturalist," "right" or "left," including those that deny that an "argument," "theory," or "explanation" is being offered in this "philosophy." For their suggestiveness only, I offer the following sample of examples and topics and a very rough indication of their ordering.

After a discussion of calculating with disappearing nuts (and immediately after a discussion of the practicality, "in certain circumstances," of counting by leaving out multiples of 3), Wittgenstein raises the question of what kind of certainty surrounds the demonstration *that a shape cannot be made to coincide with its mirror-image.*[47] Right on the heels of the moonshine, poetry, pain, and calculation passage, mentioned above, comes the distinction between game and experiment (illustrated by games which are always winnable), followed by the specter of division by zero, followed by a distinction between description and explanation, alongside sleeping and waking, and a laying to rest of demons one does not see. Next comes reiteration of a (mis)calculation technique in which linear sequences of marks are divided such that the bottom and top (or first and last) elements of each unit are counted *twice*, that is, each shares a mediator with its neighbor. Then follows Russell's paradox as a "natural" circle, the "not-blue" passage; and a more complete list of motifs in this section threatens to be global: according to my notes, "health/disease, one/many, writing/speech, dead/living, order/disorder, memory/mechanism, practice/theory, negation/contradiction, certainty/uncertainty, god/demon," and more, all in just a few pages. The motifs of exclusion, circularity, and mimesis seem nevertheless to be unavoidable bridges between Wittgenstein's mathematical and anthropological reflections as a whole.

For reasons that will shortly become clearer, my personal notes to this section of the *Remarks* also allude to the "mathematics" of *A Midsummer Night's Dream*. In certain respects, I admit, Wittgenstein's often fictive imageries seem less coherent than Shakespeare's and Beckett's staging of mathematical techniques, although his "logic" resembles theirs. He reiteratively circles the "fundamental" brink of negative mimesis where they make it central, explicitly as the Beckettian "fundament" and as Shakespeare's Bottom. Yet the "not-blue" passage, and its circular context, gives us the mimetic model we want — but only if we make it intersubjective instead of a dialogue between God and man. I thus propose to show how simple mimetic models can generate grand Wittgensteinian problems.

The simplest mimesis generates a simple iteration like Wittgenstein's circle of 2s from an arbitrary starting point (arbitrary since, by stipulation, the only law at work is mutual imitation). Now consider the negative equivalent: whatever arbitrary number you begin with, I will counter with a different number, which you will counter in turn, and so on. An apparently random sequence is thus generated so that, even if the number of ways of being different were finite, no inspection of any finite iteration could determine that the sequence was not random, although its chances of being so would apparently diminish continually and we might bring a contingent decision to bear at some point: "But suppose now [a caveman] developed π! (I mean without a general expression of the rule)."[48] In the case of an unlimited number of numbers to choose from, even a contingent or arbitrary decision would then seem "doubly" arbitrary (inasmuch as the chance of A and B repeating each other would be infinitely small even if they were *not* determined by mutual exclusion). However, "inspection" of an infinite iteration would determine the case if we defined randomness as a (meta)rule *excluding* rule, in the sense of the indefinite (infinite) iteration of a finite pattern, while *including* every possible finite pattern, namely, in terms of *exclusion of exclusion.*

Now consider the case of metadifferentiation: Given A and B in the situation above, A may decide that since their differences amount to metasimilarity (an agreement to be different), he will be *really different* by playing the *same* number as B. Then, if B does the same and claims to maintain any semblance of difference other than *order* (the invidious distinction of being second), he must claim to *do* the same but to *be* different precisely on the grounds that he is *not* attempting to be different (a mimetic crisis or galloping "fade-out" that recalls the instabilities of fashion). If B maintains his original strategy, on the other hand, the iteration will look like positive imitation by A of genuine independence on the part of B (an inversion of the truth since A has *really* made the difference), although this independence is again belied by B's never repeating A's number. Of course, if B decides to solve this difficulty by copying A at random, negative mimesis will then look like a combination of positive mimetic order (A) and genuine independence (B) even "at infinity." More provocatively, since B's iteration will then *really* be random, he will have achieved genuine independence by way of total interdependence. The point here is not simply that there are different "readings" of the same iteration (of the kind exploited by Wittgenstein's skepticism about rules), but that this situation recalls more pointedly both his suspicion that the "unordered generalities" of mathematics may, by strategies of exclusion, be masked as an order founded on positive independent principles and his view that the coincidence of mathematical semblance and reality is of a kind that classical objectivists and subjectivists alike "turn their backs on."[49]

Should this model itself seem merely "subjective," it is worth mentioning a purely numerical analogue: one where, say, the rule of exclusion of 9 from a decimal sequence is translated into a sequence of base 9. Supposing that the results of die throws are written from right to left and read after each throw as base-9 numbers, the rule is that the latter cannot be translated into decimal sequences that include 9. Now, such an infinite sequence would *not* exclude any finite sequence (our primitive definition of randomness), but would merely iterate a new rule of exclusion for every throw of the die, that is, we would have to use different dice (from which different numbers were missing) on different throws that could not be predicted, or planned in advance. I think that Wittgenstein's notorious ambivalence on the subject of the real numbers relates to this situation. For it cannot be ruled out that *all* "random sequences" (i.e., all infinite sequences) may be proved, by standard mathematical procedures, to conform to *some* rule — and it was in the light of this kind of threat to the distinction between law and arbitrariness that Wittgenstein seems to have regarded π, even superstitiously, as indistinguishable from something monstrous and unformed, generated by a new set of exclusions at every step of its expansion. Thus he expressed a rather cantankerous skepticism toward any conceivable ("Russellian") proof that some sequence X is included in π without showing where (i.e., *by excluding its exclusion*), just as he brought a parallel though inverted skepticism to Dedekind's cut, since it would work "even if the mathematics of the irrational numbers did not exist."[50]

Must "the infinite" henceforth be sacrificed? I shall return to this in connection with Beckett, in whose hands π turns out to be a dentist. For the present I will rest my mimetic case — intended as a new description rather than an explanation of Wittgenstein — with the latter's recommendation that we be satisfied, even in arithmetic (specifically, an imaginary one in which a contradiction has been discovered), with "certainty" of the rather dubious kind which holds "that in general there *won't* actually be a run on the banks by all their customers."[51] Whatever it may say about mathematics, this is at any rate consistent with what Keynes asserted about mimetic limits to the mathematical science of economics as it concerned *speculation*. For Keynes elegantly compared investment to a contest where a competitor's success depends on his ability to pick "not those faces which he himself finds prettiest, but those which he thinks likeliest to catch the fancy of the other competitors, *all of whom are looking at the problem from the same point of view.* . . . We have reached *the third degree . . . anticipating what average opinion expects the average opinion to be.* And there are . . . fourth, fifth and higher degrees."[52] Here we reach an evident merging of so-called objective and intersubjective reality, of calculation and mimesis, of the philosophy of science, social science, and literature.

4. Puck, in *A Midsummer Night's Dream*, (III.ii.437–38, 462–63), summarizes the play as follows:

> Yet but three? Come one more.
> Two of both kinds makes up four.
> . . . .
> Nought shall go ill;
> The man shall have his mare again, and all shall be well.

Actually Wittgenstein also summarized this play well when, in trying "to describe how it comes about that mathematics appears to us now as the natural history of the domain of numbers, now again as a collection of rules," he imagined studying animals by "pass[ing] transformations of animal shapes in review" — not as "a branch of zoology," but in terms of "mathematical propositions."[53] Certainly, the *Dream*'s myriad bestial transformations resemble the purely formalist transformations of structural anthropology more than the poetic evocations of nature so often celebrated by literary critics. Wittgenstein's distinction between poetry and "psychological experiment" also applies nicely here since these characters have no "internal" psychology at all, but are governed entirely by the system of transformations. (But this does not mean that the play is not a "psychological experiment" in other respects, as is suggested via the play-within-the-play, or *metaplay*. Wittgenstein's analogous distinction between prediction and proof, experiment and calculation, did not prevent him from considering proofs predictions or *metapredictions* about future proofs.)

If I had more space, I would try to illustrate how Shakespearean arithmetic provides a better context than most of Wittgenstein's own "literary" scenarios for the endless examples of systematic miscalculation he dreamed up. The *Dream* can easily accommodate, for example, the "*doubled mediator* technique" I cited earlier, where $2 + 2 = 3$, $3 + 2 = 4$, and so on. Moreover, division by zero — axiomatic in Shakespeare — is division by Bottom, the nought who has certainly gone very ill (to paraphrase Puck).[54] Bottom, actor or mimesis incarnate, can (or rather wants to) be *anything*, so the result of *dividing* by him is (almost) arbitrary, whether he be conceived of (here) as an anus or (as zero is more obviously conceived of elsewhere in Shakespeare) a vagina. At any rate, he is congruent with the "Law of the Ass" in *Much Ado About Noth-*

ing: demonstrably the fashion-law of mimetic exclusion (recalling again Wittgenstein's comparison of mathematical order to fashion, in connection particularly with the inductive proof of consistency). We are dealing here, to make the inevitable pun, with cal-*cul*-ations — and, lest the mimetic moon be forgotten, with calculations in which the Shakespearean moon more or less explicitly "bores a hole" through the earth and takes the place of the sun (III.ii.52–55). Indeed, in this startling passage the supposed murderer (Demetrius) also takes on the "dead" semblance of his supposed victim (Lysander), connecting mimesis and sacrificial logic.

I must restrict myself here to Shakespeare's fundamental arithmetic, as stated by Puck, which (unlike Wittgenstein's) *does* correspond to our mathematics: $2 + 2 = 4$, $3 + 1 = 4$. What Shakespeare added here does not question the law of the excluded middle, as Wittgenstein did, but merely reiterates the law of the *excluded bottom.* The essential thing to note is as follows: just as Puck's "nought" represents the foundational *inclusion of the excluded* in the realm of human and divine eroticism (thus Titania loves the ass-Bottom, animal-man, erotic-zero, butt-head), so Puck turns the innocent orderings of arithmetic into laws of exclusion: 2×2 *kinds* $= 4$. (Beckett's Saposcat — or perhaps *Homo scatologicus* — similarly finds all calculation idle except "concrete numbers" in which the unit is specified.[55]) What Puck most obviously means here is male and female, the most banal of such laws. But the two kinds also refer to a more general bifurcation whereby the 3 is also distinguished from the 1, as the loved or lovable is distinguished from the unloved or unlovable — since it is for the now-unloved Hermia that Puck is waiting. Thus the 1 (which numbers Helena and Hermia successively) twice plays a part identical to 0 (the excluded Bottom of love). Shakespeare gives us three main patterns for his four central lovers prior to their "happy" marriages: Helena (L) Demetrius (L) Hermia (L) Lysander (L) Hermia,

$H.1$ (L) D (L) $H.2$ (L) L (L) $H.1$

— an unrequited CIRCLE —

$H.2$ (L) L (L) $H.1$ (L) D (L) $H.1$,

where (L) is the operator "loves," and $H.1$ and $H.2$ are Helena and Hermia, respectively.

Now, granting only mimesis and chance (and excluding homosexuality, or more "bottoms") as operators of the fourfold combination (a Wittgensteinian overdetermined "four-footed" cal*cul*ation if ever there was one), the first and third patterns (which are symmetrical in terms of the simple interchange of the two *Hs*, so their equivalence implies *H*.1 = *H*.2) can only be generated by a combination of positive (= male) and negative (= female) mimesis. (These latter equations, besides being proverbial, follow from o = vagina elsewhere.) Moreover, the equivalence of the two patterns simply establishes that the "Top Model" (def: loved by more than one person) is equivalent to the "Bottom Model" (def: excluded from the set of the loved), just as Bottom himself is elevated to Titania's love. The "paragon-paramour" is "a thing of naught" (IV.ii).

It remains only to account for the second pattern, the circle of unrequital between what Girard would call the mimetic masochists.[56] But this can be done even more simply (at least in principle) by appeal to the model of negative mimesis in the previous section. There is one complication, however: the operator "loves" must now be taken as identical with the operator "imitates" (i.e., mimesis), using the identity paramour = paragon above. (In "Enough," Beckett also explores just such an identity.) Then, by iterating the final model of negative mimesis given in the previous section on Wittgenstein, we arrive at rock bottom, the Shakespearean circle:

$$A \ (I) \ B \ (I) \ C \ (I) \ldots (I) \ Z \ (I) \ A,$$

where *I* is the mimetic operator, and each couple behaves like *A* and *B* in my model (*A* imitates *B*, who imitates independence from *A*, i.e., *C*, etc.) *Q.E.D.* That such independence is really a mask for mimetic arbitrariness is confirmed in the *Sonnets* by love "at random from the truth." Shakespeare explicitly equates love with order or form founded on disorder, the *deformed*.

I return, finally, from moon to earth by stressing a significant homology here with nonclassical logic. Reichenbach's three-valued logic, for example, entails two kinds of negation, "complete" and "cyclical," the latter of which operates exactly like the circle of exclusion: $A \neq \text{not-}A \neq \text{not-not-}A \neq \text{not-not-not-}A \ (= A) \neq \text{not-}A$, etc. Moreover, Reichenbach developed this logic specifically to describe quantum mechanics, the *physics of cognitive exclusion*.[57] Such logics are no more or less necessarily "arbitrary," I claim, than my simple mimetic models.

5. Beckett will be even more pitiably mutilated here than his predecessors were above. I can barely summarize, for example, how he illuminates Wittgenstein's "moonshine" analogy between calculation and pain (not to mention poetry), so I will simply note two essentials.

First, Beckett's Malone explicitly identifies the rational construction —
"It is thus that man distinguishes himself from the ape" — of the Object ("shorn of all its accidents") with pain and loss so that the exclusion of pain is formally a double exclusion, that is, the exclusion of
the image of the painful Object, which itself excludes arbitrariness.[58]
Second, Beckett's "logic of pain" is generally consistent with Girard's
"masochist induction," with its simplest reasoning to the effect that, for
example, since I have lost a series of loves to my mimetic rivals, I must
proverbially love only what I can't have. In its proverbial form, the Girardian induction appears to be deduction (from human nature), wholly
excluding arbitrariness in favor of "four-footed" Wittgensteinian calculation. Thus pain "is not a form of behaviour" here, just as calculation is
not an experiment, precisely inasmuch as it appears to be a self-fulfilling
prophecy. (In the simplest terms of Shakespeare's *Dream*, "true" love
authenticates "true" pain, and vice versa.) Pain is indeed generally considered by philosophers an instance of epistemological indubitability
in which the subjective and the objective coincide. The trick, then, is
to map this coincidence onto a model where pain is *not a given*, but a
product of a circular "loop" in which it appears successively as cause
and effect — or, as Beckett put it, emphasizing the mimetic element, a
model of "speculative pain." But I must sacrifice this line of inquiry,
since there is only enough space here to briefly juxtapose Beckett's π
with Wittgenstein's.

As though by magic, Wittgenstein actually described the basic plot of
Beckett's *Molloy* even more accurately than he did Shakespeare's *Dream*,
in his description of a language game that uses "reports, orders and so
on," and in which "people . . . say [calculating propositions] to themselves perhaps, in between the orders and the reports."[59] This is exactly
how part 2 of *Molloy* is structured, that is, as a narrative of what happens
in between receiving orders and writing a report, in which Beckett's
calculations with π occur. Four basic axioms of the work, which share
much with Shakespeare and also recall the Dodgson puzzle, can be
summarized as follows:

1. The Ass-Anus = "Nothing" = the FUNDAMENTAL confusion of
"fundamentals and accidentals," of deduction and induction. Law of
Exclusion (Law of Law itself) (1).

2. The model of identity construction is negative mimesis, that is,
negation of similarity (accomplished in the text by a simple act of
murder, since the dead do not resemble the living). Law of (Sacrificial)
Exclusion (2).

3. Narrative truth is twice structured circularly: in part 1 as an *in-*

complete circle (defined by an "arbitrary" narrative discontinuity), and in part 2 as a complete circle *whose completion coincides with self-contradiction*, or overt lie. (The lie is "it is midnight," or, precisely à la Dodgson, that one day is ending and another beginning.) Law of (Epistemological) Exclusion (3).

4. The *relation* between false terms "is not necessarily false, so far as I know." (Mutual Inclusion of True and False.)

The most obvious calculation in part 1 (apart from its calculation of farts per minute) produces Molloy's celebrated system for sucking all of his stones in a sequence such that he never sucks the same stone twice in the same cycle. Solution: leave one of his four pockets of stones *empty* (i.e., 3 + 1, where "1" contains 0). While this problem is genuinely "mathematical" in excluding overt reference to subjectivity and in producing cyclical iterations, Molloy nevertheless exhibits a Wittgensteinian anxiety vis-à-vis the "utter confusion . . . bound to reign . . . [in] the sum of all cycles, though they went on forever"—a confusion that will shortly be linked to pain.[60] It is not entirely surprising, then, that the introduction of π in part 2 is embedded in an intersubjective identity crisis between Moran and his son—a mimetic crisis defined precisely *in terms of each one's belief in the other's pains*—or that "Py" is the name of a person, a dentist.

Dentists are first mentioned in connection with Molloy's fart calculations, made while he is dressed in the *Times Literary Supplement* (a grotesque summary of the Beckettian relation between literature and mathematics). Here Molloy makes a mock attempt to justify his narrative inclusion of farting (the excluded), "realistically," by calculating its frequency—an activity which, we are told, "is like one dying of cancer obliged to consult his dentist."[61] This is comprehensible because Py *is* a dentist and because we know that no mere calculation of positive frequency can determine the relation between order and disorder, inclusion and exclusion—except (perhaps) "at infinity," hardly a sanguine prospect for one dying of cancer.

Meanwhile, "toothache" (you will have to take my word for it)[62] is one name for the sacrificial disease itself: the disease of sacrificial exclusion-internalization as primitively instantiated by eating. Py is the dentist because π appears to offer the promise of—precisely what Wittgenstein feared as mathematical or metamathematical lunacy—an infinite iteration of exclusions that coincides with a unique law of inclusion, the abolition of arbitrariness. Moran's son, Jacques (a scientist), claims that his toothache has *not* been cured by Py, while Moran claims to have heard from Py himself that Jacques must no longer be in pain since the tooth has been successfully "dressed." But, more generally, Py is of the opinion that Jacques "was born with bad teeth . . . and all his life he will have bad teeth. Naturally I shall do what I can. Meaning, I was born with the disposition to do all I can, all my life I shall do all I can, necessarily."[63] (Note particularly that doing *everything possible, necessarily* is the definition of an infinite arbitrary sequence, here syntactically parallel to the disposition to feel pain itself. In reconsidering my earlier description of the cause/effect "loop" of "speculative pain" — a loop considered by Malone specifically in terms of the relation between guilt and pain — one might recall Leibniz's view of the infinite iteration of causes as a paradox of the *necessity of contingency*.) To his son's request to see Py again, Moran retorts:

> Mr Py is not the unique dentist of the northern hemisphere. I added rashly, We are not going into the wilderness. But he's a very good dentist, he said. All dentists are alike, I said. I could have told him to get the hell out of that with his dentist, but no, I reasoned gently with him, I spoke with him as an equal.[64]

Circularity incarnate, Py is the transcendental emblem of undifferentiation between fundament and accident, the formal indistinguishability of true and false transcendence. But if all "irrational" dentists are alike, "rational equals," as Moran says of his scientific son (in a way that might again recall my Fregean epigraph), shouldn't at least one of them (by axioms 2 and 3) — Py himself — be sacrificed? Or should we, contra Wittgenstein, put our faith in the nontranscendent rationality (at least) of the Transcendental Irrationals?

As a matter of fact, having been cruelly compelled to sacrifice his appointment with Py, Moran's son shortly gets another sacrificial eating disorder (stomachache) from eating shepherd's *pie*, which Moran claims to cure with an enema. Sheep are the mimetic-sacrificial animals par excellence in Beckett,[65] the means by which he conjoins in "*pis*" two transcendental mediators of identity difference — the irrationals and the Christian Trinitarian equations: God = Man, God ≠ Man, and all men are "equal" in the sight of God. (Wittgenstein himself compared the propositions "the class of lions is a lion" and "God created man."[66]). In short, Beckett treats intersubjective "equality" and mathematical differentiation as aspects of the same system — a sacred one in which the excluded or transcendent contradiction is *literally* internalized, eucharistically, by being eaten or otherwise dentally manipulated. (Malone's Moll only narrowly escapes swallowing her loose tooth, which has been drilled in the shape of a crucifix.)

Is there a moral here for Wittgensteinians, left, right, or center? If so, it may be that Wittgenstein's relentless ambivalence regarding his celebrated sacrifice of the excluded middle in relation to π (i.e., his sacrifice of the real numbers "as such") followed something like the path we might deduce from Beckett. It is perhaps a matter of judgment, even of *taste*, to what extent one either reads open-mindedness or originality into this ambivalence or merely hears a new de-formation of the old pharmakon-ambivalence characteristic of sacrificers. I have argued that, inasmuch as this mimetic-sacrificial ambivalence also dominated Wittgenstein's *double* project in philosophy at large, we may, perhaps uncharitably, tend toward the latter alternative. But this can also be stated as praise.

It may be, in short, that we prize Wittgenstein's semi-concealed mimetic logic in all its literary rigor—"literary" not only in the colloquial sense that might be applied to his style, but also in the Beckettian sense of *excluding fundamental disciplinary exclusion*—over any so-called objectivist, culturalist, or subjectivist views of mathematics itself that his often (very) antagonistic admirers propose to distinguish or to promulgate. It is immediately after referring to the moon's arse, I might add, that Molloy demonstrates how all of the sciences, but above all anthropology, are constituted by relations of exclusion—"as though [man] were no better than God."[67]

What conclusions can be drawn from all this, in particular from the dominance and virulence of exclusive relations which proved as catastrophic to the objectivized project of a Frege as they seemed, perhaps predictably, ubiquitous elsewhere? Can we conclude that both so-called objective and subjective constructions of "rationality" are naturally founded on relations of exclusion, but also that these relations seem to indicate the moonlit horizon—if not a Nietzschean murder at high noon—of "rationality" "itself"? If we allow Wittgenstein or Frege any "real or philosophical rigour" on the subject of number, will this necessarily exclude Beckett and Shakespeare from the same circle? Or does the exclusion of literature from "science" or "philosophy" simply and not-so-simply obfuscate, as well as frankly beg the question of, the meaning of the overlapping patterns I have described here?

Various games going under the name of "Prisoner's Dilemmas" will allow us finally to pose the question of "intersubjective rationality" more epigraphically, and so as to illuminate once again both mimesis and its relation to infinitist questions. Indeed, most classic versions of the dilemma provide exemplary mimetic crises. For example, two players may either cooperate (C) with or defect (D) on one another: (C)(D) is scored ($0)($3), with the "sucker's payoff" for (C) — the cooperator; (C)(C) gives ($2)($2), while (D)(D) gives ($1)($1). The classic "rational" solution to this simple dilemma is that since *whatever the other player does, you are better off defecting*, you should "rationally" defect. Clearly, "rationality" here depends on the hypothesis that the players are genuinely independent of each other, neither both telepathically mimetic nor both culturally subject to the charitable but self-sacrificial "delusion" that unilateral cooperation is always to be preferred.

Of course, if the object is merely to win, to gain a *relative* advantage over one's opponent, it matters not at all whether both players cooperate or both defect. The point is that in *absolute* terms (supposing these to be measurable in dollars) the "rational" solution is mutually (sado)masochistic and clearly based on the hypothesis of genuine independence, or the *exclusion* of an "irrational" mimetic hypothesis. The irony to savor (which also recalls Girard's "masochist induction") is that two "rationalists" will *in fact* do the same thing precisely on the basis of their mutual individualistic "difference," thus the *insane* hypothesis that consciousnesses do not in fact exclude one another appears to be the preferable ground of unanimity. I have discussed elsewhere the Kantian delusion, the categorical imperative that demands unanimity among all "rational beings."[68] Here a simple game (one whose influence, however, has filtered deeply into geopolitical and economic theory) suggests that "irrational" unanimity may be more "reasonable" than its "rational" counterpart.

The iterated and inductive form of the game has proved inspiring to evolutionists, among others, since it seems to solve the dilemma (and, for some, the very "paradox" of unselfish behavior) in terms of the players being able to rationally base "cooperative strategies" on their mutual experience. Computer buffs (playing large evolutionary "seas" of game programs against each other) have demonstrated how cooperative strategies tend to survive in various simulations of similar games. (The remarkable and apparently almost totally unexpected success of the *purely mimetic* strategy [TIT-FOR-TAT] in such experiments is quasi-legendary in some accounts of game-theory history.[69]) Yet reasoning derived from indefinite iteration, however suggestive it may be for the survival of populations and genes, has rarely satisfied "rationalist" puzzlers. In particular, the iterated form of the dilemma I began with can easily be stated provided that the number of game-encounters is finite and known in advance: since it is "rational" to defect at the *last* encounter (when one's opponent cannot retaliate), one should defect at the next-to-last encounter, arriving by induction at the original single-time dilemma. (This reasoning, inductive in the logical sense, has also been variously instantiated by experiments involving human players, inductive in a psychological sense.) It seems that only an Optimist/Rationalist of the kind that Nietzsche's rather comic *Übermensch* most despised could deduce from human history that we have been genetically programmed "to cooperate." (Even if we were, *reasoning* has certainly put an end to all that!)

Michel Serres has discussed, in the context of Descartes and Leibniz, the connection between the goal of transcendent rationality as such and the goal of *always winning* in mathematical game theory, just as Wittgenstein distinguished both games and calculations from experiments. Jacques Lacan has popularized the psychoanalytic significance of prisoner-type dilemmas, while Jean-Pierre Dupuy has tried to reinterpret in Girardian terms the Lacanian significance of infinitist game theory. As originally presented by the French mathematician Jean-Michel Lasry, a significant consequence of infinitist "winnability," in a two-person game governed by the infinitely iterated specular interplay of mutual (meta)hypotheses, is that it makes plausible the transcendence of Lacan's *symbolic* domain (which establishes an independent symbolic order) over the *imaginary*, the domain of specular subjectivism. Dupuy would polemically reverse this, making the symbolic merely "a *product* of the imaginary, or, in other words, of mimesis indefinitely reduplicated."[70] Yet neither position addresses the potential mathematical crisis provoked in game theory by the contradiction between the axioms of *determinacy* and of *choice*, the first of which is necessary to infinitist "winnability," while denial of the second, as Gödel showed, leads to contradictions in finite mathematics. Since we are concerned with metaphors of winning, "determinacy" and "choice" translate aptly into ordinary language, although they are defined mathematically by the absence or presence of one-to-one correspondence between two infinite sets (a power set and a set of Turing degrees) — a correspondence whose *finite* equivalent, of course, entails being *even*. It may well be, then, that neither Lasry nor Dupuy makes his case, since both mathematical and psychoanalytic "determinacy" and "choice" remain open to, indeed *compel*, interpretation. From this point of view, in fact, Wittgenstein's comparison of the meaning of the axiom of choice with the equation $0 \times 0 = 0$ may well be more suggestive than any rejection (or embrace) of infinitism "as such."[71]

We have come full circle to 0 and Bottom. Furthermore, any *finite* version of the Dupuy-Lasry game would entail both players making *false inductions* from an arbitrary "bottom" or zero point which together may be truth-productive and thus winning — like Beckett's true relations between false terms and Shakespeare's Bottom, whose dream *has no bottom*. Such a simple specular model of knowledge production from falsity and ignorance might appear arbitrary; yet surely we may generalize that any truly evolutionary approach to knowledge will, at some point, have to posit *some* plausible model of the paradoxical generation of knowledge from total, undifferentiated ignorance — however and wherever this may be located in the passage from things to animals to philosophers.

I thus conclude with an allusion to Serres's summary of Leibniz's arguments against Plato's, Descartes's and Locke's constructions of knowledge — from reminiscence, reflection, and the zero point (tabula rasa), respectively — since in each case reduplicated *iteration* is the weapon used.[72] In the case of reflection, Leibniz argued that it is only "by ellipse," with its *double* center, that such an act seems primitive; and, in the case of any attempt to construct the genesis of knowledge "from zero and the exterior," he argued that contingent understanding never in fact constructs from zero, but rather by "reconstituting the plus from the minus."[73] Serres claims, against what he calls the common view, that for Leibniz this did not mean *excluding* the zero construction or projection, "as one cuts the false from the true," but rather conceiving of one in terms of the other, "a little as the continuous invades the discontinuous, glueing [*en gommant*] the vacuities."[74] (Accordingly, against a common contemporary view, I have not tended to glorify Wittgenstein at Frege's expense.)

Husserl also claimed, like Descartes, that our knowledge of a priori "segregation" — as opposed to that identity differentiation whose "ulterior purpose" is merely to distinguish "similars" — is only to be known through *reflection*. This zero state of knowledge, a form void of content, may indeed be "objectively primitive" (like Frege's summary of all falsehoods in zero, the universal mediation of their similarity), yet in order to account for the production of objectivity we must also, according to Beckett (and Girard), conceive Husserl backwards, that is, from the production of segregation via the mutual "reflection" of the similars, the primitive reflectors themselves. Thus we come again to the suggestiveness of Leibniz, for whom the zero states of Locke's "savage" — mere forms waiting to be written on — were "states of decadence," savages being not really "primitives but men who have forgotten the primitive."[75]

No one doubts, of course, that literature, philosophy, and mathematics once had "primitive" connections. I have merely tried to show that their more contemporary objects and "objectivities" are, like all that seems prime, no less compellingly entwined. Such a task would be superfluous if one did not still often read, for example, that literature differs from mathematics by being "simply a product of opinion . . . subject to permanent disagreement like the ideas of literary criticism" — and this from two mathematicians who present themselves as philosophically very broadminded when it comes to the epistemology of mathematics.[76] But Serres's view that literature is "a generalized physics" is obviously far from being generally understood even by very sophisticated people, let alone accepted (which is another matter).

NOTES

In memory of E. T. W. Smyth, mathematician (in expiation of the mathematical lacunae inhabited by his son).

1   For an introduction to Girard's thought, see his *Violence and the Sacred*, trans. Patrick Gregory (Baltimore and London, 1977).

2   This is quantitatively true of all except Wittgenstein, whose single moonrise in *Remarks on the Foundations of Mathematics*, however, is qualitatively spectacular.

3   Paul de Man used this phrase in a conversation with me about Wittgenstein. It may also appear somewhere in his work.

4   I note a parallel here with, say, James Feibleman's interpretation of the binary operations of Boolean algebra as "similarities and differences," and his claim that "in postulating logic as the basis of mathematics, we have learned the greater generality of difference [over identity]"; see his *Inside the Great Mirror: A Critical Examination of Russell, Wittgenstein and Their Followers* (The Hague, 1973), 190. Similarly, in connection with the ambivalence of Russell's attempt to reject his own early realism: "He has admitted that he can get rid of any universal except similarity, which cannot be explained away, like 'or' or 'not,' as belonging only to speech" (27).

5   Samuel Beckett, *Molloy, Malone Dies, the Unnamable* (New York, 1958), 82.

6   I would identify this with what Barbara Herrnstein Smith suggests is the epistemological symmetry between true and false theories, granting that this puts in question the very definition and project of epistemology; see, for example, her *Contingencies of Value: Alternative Perspectives for Critical Theory* (Cambridge, MA, 1988), 123–24. But this putting in question is, *globally*, "epistemology" itself, and it does not preclude the pursuit of *local* assymmetries. Local epistemology is necessary precisely insofar as global epistemology is uncircumnavigable.

7   My book, *The Habit of Lying (Or Fundaments of Fiction)* (forthcoming, Duke University Press), explores this topic in depth.

8   In French "point selle" allows a scatological pun appropriate to Shakespeare's and Beckett's view of such points of equilibrium: absolute kings gave audiences "en selle"/in the saddle, that is, on the privy.

9   Michel Serres, *Les Origines de la géométrie* (Paris, 1993), 195–291. Perhaps for reasons of intellectual politics, Girard is mentioned only once by name, although his encounter with Girard now clearly dominates Serres's whole project in the history

of science. (The contemporary age may provide instances of the opposite sacrifice to Euclid's, that of "pure" geometry in favor of algebra; something like this has been claimed by René Thom and others.) For an interesting discussion of connections between Thom and Girard, see Lucien Seubla, "Vers Une Anthropologie Morphogénétique: Violence Fondatrice et Théorie des Singularités," *Le Débat* 77 (1993): 102–20.

10 John Bigelow, *The Reality of Numbers* (Oxford, 1988), 93.

11 Edmund Husserl, "On the Concept of Number: Psychological Analyses," trans. Dallas Willard, *Philosophia Mathematica* 9 (1972): 44–52; and 10 (1973): 37–87; quotation on 70.

12 However, retrospection and temporality as well as representation are sites of an acute "identity-crisis" in Husserl's essay: "We can relationally unite past representations neither with each other, nor with present representation" (ibid., 48). Although the moon usually dominates his "arbitrary" illustrations of numerical collections, Husserl's surprising illustration of the number 2 by "the clock and the pen" is, I think, a rather fateful summary of the concealed relation (mediated by temporality and the double meaning of "priority") between falsehood and iteration which the essay steadfastly circumlocutes. Frege, like Husserl, vehemently rejected the Kantian identification of number with both similarity and temporality. To identify number with time, he wrote, is to "number number." Frege implicated Leibniz, and Husserl implicated Hamilton, in this temporal "delusion."

13 Gottlob Frege, *The Foundations of Arithmetic*, trans. J. L. Austin (Oxford, 1989), x; Ludwig Wittgenstein, *Remarks on the Foundations of Mathematics*, ed. G. H. von Wright, R. Rees, and G. E. M. Anscombe; trans. G. E. M. Anscombe (Oxford, 1989), 433.

14 Cf. Feibleman, *Inside the Great Mirror*, 25.

15 Frege, *Foundations*, 65.

16 Ibid.; my emphasis. Besides thinking of Frege's comparison of the moon with almost everything in the *Foundations*, I am also alluding here to his assertion that "an a priori error is thus as complete nonsense as, say, a blue concept"(3). The stress on exceptionality here is in any case Frege's own.

17 Ibid., 101.

18 Flann O'Brien, fine novelist of number, confirms this in *The Third Policeman* (1967), where the highest law (synthesis of mathematical and nonmathematical order) is contained in a room inside a wall. (The wallpaper on the *inside* of this wall reminds me of Wittgenstein's imagined use of mathematical equations as wallpaper in the *Remarks*.) O'Brien's other masterpiece, *At Swim-Two-Birds* (1939), is entirely structured on the relation between the even and odd numbers, recalling Serres's treatment mentioned above.

19 Brian Rotman, by contrast, insists on the metalevel status of zero, linking it historically to that of the perspectival vanishing point in painting as well as to semiotic developments in economics; see his *Signifying Nothing: The Semiotics of Zero* (Stanford, 1993 [London, 1987]).

20 Frege, *Foundations*, 87ff.

21 Ibid., 81.

22 Ibid., 94.

23 However, as noted by his translator Dallas Willard, Husserl himself later traced the "childlike" naiveté of his early essay to his ambiguous use there of the term

"representation." As regards set theory (and keeping mimetic rivalry in mind), see Bigelow's description of *n*-membered sets as "rivals" to the fundamental title "*n*," the consequence of the impossibility of defining *n* in terms of the set of all *n*-membered sets, in *Reality of Numbers,* 56. Bigelow compares von Neumann's positing of an *n*-membered set which all others *resemble* to Bishop Berkeley's Model Horse (which all real horses "resemble"), noting further that this procedure entails supposing a "proper class" which is a member of no set or class and has no "properties" except universal one-to-one pairing (46–47). See also Nelson Goodman, who defines representation itself in terms of one-to-one pairing in a continuum, in *Languages of Art* (Indianapolis and Cambridge, 1976), 225ff.

24 Ludwig Wittgenstein, *Tractatus Logico-Philosophicus,* trans. D. F. Pean and B. F. McGuinness (London and New York, 1961), § 6.522, § 6.53.

25 Wittgenstein, *Remarks,* 291.

26 Ibid., 256; see also *Tractatus,* § 6.54.

27 Girard's scapegoat, first expelled from the community under the sign of evil, always returns under the sign of transcendent divinity (because of the peace which his exclusion has brought); see, for example, *Violence and the Sacred,* 1–68.

28 Wittgenstein, *Remarks,* 422.

29 Ibid., 282.

30 Ibid., 369.

31 Ibid., 406, 214.

32 Ibid., 243. I am keenly aware that this entire passage (IV.35) may *appear* to mean the opposite, since it begins, "Being educated in a technique, we are also educated to have a way of looking at the matter which is just as firmly rooted as that technique." The point, I believe, is nevertheless that *both* "objectivists" and "subjectivists" acknowledge the fundamental character of mathematical "semblance" by "look[ing] away . . . at something else," an interpretation that is confirmed by the passage on calculation, poetry, and pain (201–2). See also Beckett, *Malone Dies,* where Malone's mother declares that the sky "is precisely as far away as it appears to be" (268) — a sky about which Malone takes strenuous precautions before turning his back on it (190).

33 Wittgenstein, *Remarks,* 344, 242, 18 (this last term is in the table of contents provided by the editors).

34 Ibid., 392, 405, 197.

35 Ibid., 345.

36 Ibid., 221, 351, 373.

37 Ibid., 201–2.

38 Ibid., 227.

39 Ibid., 404.

40 Ibid., 415. Cf. Italo Svevo's *Confessions of Zeno* (1923), which is defined by mimetic substitutions throughout and dominated by the iterative motif "my last cigarette." I am also reminded, regarding the relation between first-level and metalevel iteration, of Brian Rotman's linkage between the kind of metalevel "fade-out" that characterizes successive iterations of exponentiation (addition, multiplication, exponentiation, hyper-exponentiation, etc.) and ordinary numerical iteration. Rotman's anti-infinitist arguments accord with Wittgenstein's vehement rejection of his earlier logical "idealism." However, Rotman's imaginatively "physicalist" solution to the problem of first-level iteration is a non-Wittgensteinian one that I suspect would have pro-

voked Wittgenstein to question the *relation* between the physicalist and mentalist and/or semiotic constraints governing Rotman's finitist arithmetic first-level and exponential fade-out, respectively. See his *Ad Infinitum . . . The Ghost in Turing's Machine: Taking God Out of Mathematics and Putting the Body Back In* (Stanford, 1993), especially chapter 4.

41 Wittgenstein claimed that his task was to attack (e.g., Russell) "from without," namely, from the privileged site of the excluded, the "philosophic"; see *Remarks*, 383.

We should recall that the *Tractatus* articulated a doctrine of the unity of ethics and aesthetics (see § 6.421). It may also be worth recalling, in view of the early Wittgenstein's well-known interest in Kierkegaard, that the latter regarded the ethical realm as the realm of mimesis (bordering on the religious realm of sacrifice). See, for example, my *Question of Eros: Irony in Sterne, Kierkegaard and Barthes* (Gainesville, 1986), 371.

42 Wittgenstein, *Remarks*, 316.

43 Ibid., 207, 102. Elsewhere (105), a primitive logic in which double negation is identical to single negation is compared to a number system whose iteration stops at 5. Negation (whose metalevel status was at stake in the *Tractatus* as well as here) once more conjoins the problems of bottom-level and metalevel iteration.

44 Ibid., 207.

45 Ibid., 208.

46 Ibid., 352. Primitive "power structures," of course, often exhibit precisely such patterns, although Wittgenstein's examples are loose and fictive.

47 Ibid., 92.

48 Ibid., 344.

49 Recall Wittgenstein's repeated insistence on the difficulty of distinguishing between "calculating wrong" and "not calculating," that is, between arbitrariness regarded as error and as exclusion-independence.

50 Wittgenstein, *Remarks*, 286. I have discussed J. A. Bernadete's critique of Wittgenstein's finitism in a section of my *Question of Eros* entitled "Wittgenstein's Sister."

51 Wittgenstein, *Remarks*, 401.

52 As quoted by the Girardian Jean-Pierre Dupuy, "Self-Reference in Literature," *Poetics* 18 (1989): 509.

53 Wittgenstein, *Remarks*, 230–31.

54 See Rotman, *Signifying Nothing*, for a partial analysis of this metasickness in *King Lear*.

55 Beckett, *Malone Dies*, 187.

56 René Girard devotes many chapters to this play in *A Theatre of Envy: William Shakespeare* (Oxford and New York, 1991), chapters 3–8, 19, and 27. His reading of *Dream* is mostly consistent with my own, but to demonstrate the "independent corroboration" demanded by science, I should note that I came to these conclusions before reading his book.

57 Cf. Kurt Hubner, *Critique of Scientific Reason*, trans. Paul and Hollis Dixon (Chicago, 1983), 90ff. In this volume, see also Arkady Plotnitsky, "Complementarity, Idealization, and the Limits of Classical Conceptions of Reality"; he makes a spirited plea here for taking quantum exclusions as a philosophic model, and this leads him in a different direction from Rotman's philosophic tendency to exclude infinitism; and Andrew Pickering, "Concepts and the Mangle of Practice: Constructing Quaternions"; he also treats an interesting case of exclusion, highlighting the negotiations

that led William Rowan Hamilton to abandon his project to extend imaginary numbers into *three* dimensions in favor of the leap from *two* to *four*. My own approach stresses elementary relations of exclusion that may or may not be more "primitive" than these developed ones, but which seem, for better or worse, more ubiquitous.

58 Beckett, *Malone Dies*, 254.

59 Wittgenstein, *Remarks*, 392.

60 Beckett, *Molloy*, 73–74.

61 Ibid., 32.

62 I don't know if Beckett ever read Flann O'Brien, but their symbolisms of dentists, teeth, and mimetic bi-cycles are mutually corroborating.

63 Beckett, *Molloy*, 111.

64 Ibid., 103.

65 They play the same mimetic-sacrificial role in the films of Peter Greenaway, particularly *The Draughtsman's Contract* and *Drowning By Numbers* (where sheep are also numerical). Dodgson/Carroll's *Through the Looking-Glass and What Alice Found There*, quite as mimetic as its title suggests, turns the White Queen into a sacrificial sheep.

66 Wittgenstein, *Remarks*, 403.

67 Beckett, *Molloy*, 39.

68 See Smyth, *Habit of Lying*.

69 See William Poundstone, *Prisoner's Dilemma* (Oxford, 1993); and Douglas Hofstadter, *Metamagical Themas* (London, 1986).

70 Dupuy, "Self-Reference in Literature," 513.

71 Wittgenstein, *Remarks*, 283.

72 Michel Serres, *Le système de Leibniz* (Paris, 1982), 795ff.

73 Ibid., 804, 799.

74 Ibid.

75 Ibid.

76 I quote from P. Davis and R. Hersh, *The Mathematical Experience* (Boston, 1987), but without intending to scapegoat these authors for a crime whose authorship is common.

Microdynamics of Incommensurability:

Philosophy of Science Meets Science Studies

Barbara Herrnstein Smith

A number of themes and issues encountered elsewhere in this volume recur in the following examination of what could be called the dynamics of nonconvergence. The immediate focus here is the recurrent failure of intellectual engagement in encounters between traditional philosophy of science and the critiques and alternative views of theorists, historians, and sociologists working in the relatively new field of "science studies." I am also concerned, however, with the more general theme and issue of incommensurability, which figures centrally and by no means incidentally in the debates just mentioned. The question is whether, as is traditionally maintained, rival theories are always ultimately measurable against a common standard of truth so that, at least in principle, their divergent claims may be compared and the superior ones chosen accordingly; *or* if, as is argued by a number of philosophers and historians of science, there are conditions under which supposedly conflicting theories cannot be measured or compared that way: when, for example, they assume radically divergent but (arguably) equally credible conceptions of the universe, or, as in the case of these epistemological debates themselves, when part of what divides the parties is how to conceive the standards (*truth, rationality, evidence,* and so forth) by which the merits of their divergent theories could be measured — if, indeed, merits, measurements, or even choices, as classically conceived, are relevant to the outcomes of such conflicts (if, indeed, those divergences need be seen as conflicts).

As this latter example suggests, the situation that concerns us here has a distinctly reflexive quality: that is, certain, evidently rival views of knowledge and science differ on, among other things, how to describe, explain, compare, and assess rival views. The reduplicative — echoing, mirroring — structure of this situation is intriguing and, I think, instructive. For it indicates the order of perplexity involved in these encounters and failed engagements, and it also raises the more general question — *very* general, I would say, with ethical and political resonances as well as extensive theoretical implications — of how to understand the cognitive intractability (or, as it may be seen, the blindness, stubbornness, and folly) of those who disagree with us.

In exploring these questions here, I focus on the epistemological debate as played out in the pages of a recently published book by the Anglo-American philosopher of science Philip Kitcher.[1] In a move quite familiar in these debates, Kitcher seeks, in the name of a "middle way" between two allegedly "extreme" positions — one a familiar and more or less established account, the other a relatively novel and currently controversial alternative — to redeem precisely the former: in this case, the set of interrelated ideas about science shared over the past century or so by most academic philosophers, many scientists, and much of the educated public.[2] It is to this set of familiar ideas, currently challenged in a number of quarters, that Kitcher alludes in the "legend" of his subtitle, and, indeed, he does display a certain ambivalence or affectionate irony toward some of its hoarier features. Nevertheless, his clear and intermittently explicit aim is to rehabilitate and ultimately to reaffirm the established account in all its crucial elements.

Citing the "old-fashioned virtues" of the "broadly realist" conception of science he defends, Kitcher rehearses those elements as follows: "[S]cientists find out things about a world that is independent of human cognition; they advance true statements, use concepts that conform to natural divisions, [and] develop schemata that capture objective dependencies."[3] His defense of this set of ideas could claim its own virtues as well. Unlike other defenders of besieged orthodoxy who snipe and snort at often unnamed, commonly unquoted, and largely unread "postmodernists," Kitcher identifies his adversaries explicitly, quotes from their texts directly, gives evidence of having read them in some sense carefully, and frames his objections politely and painstakingly. Moreover, he sets forth his own views of science through patient rehearsals of standard arguments, detailed reconstructions of classic cases in the history of science, elaborate analogies and thought experiments, and established models drawn from other fields, including economics and evolutionary biology.

These are substantial virtues from most perspectives and, in Kitcher's own understanding of intellectual history, decisive ones. That is, they embody and exhibit what he sees as the processes and strategies that distinguish what he calls well-designed and properly activated cognitive propensities from what he calls dysfunctional or improperly activated ones and which, accordingly, yield propositions "likely to prove true" rather than false.[4] Given the epistemic criteria that Kitcher promotes and seeks to satisfy in his book and the cognitive procedures he describes and seeks to exemplify there, *The Advancement of Science* should carry the day in competition with the more skeptical, revisionist accounts of science he seeks to refute. And, indeed, as indicated by appreciative reviews in the *New York Times* and elsewhere, it does carry the day for a number of readers — especially, it appears, those who already grant the decisive authority of those epistemic criteria and the propriety of those pointedly rational procedures.[5] As indicated by more critical reviews in other journals, however, it does not carry the day for all readers, especially not, it appears, those already persuaded (or, as it may be seen, seduced or deluded) by more skeptical revisionist accounts.[6] This divergence of *critical judgments* regarding the success of Kitcher's efforts could be explained in various ways, but the explanations of that divergence would themselves probably diverge in more or less strict accord with the tenor of those judgments. This familiar self-doubling, self-confirming regress of judgment and justification recalls the reflexive echo I noted a moment ago in traditional and revisionist views of the commensurability of traditional and revisionist views. It thereby exemplifies as well the more general structure of cognitive and rhetorical *circularity* which, as I shall suggest, is a crucial feature of the dynamics that concern us here.

Among the critiques of, and alternatives to, traditional realist/rationalist philosophy of science that Kitcher seeks to defuse or rebut, some are clearly more provocative for him than others. Thus, while he takes issue on some points with other philosophers (notably, Larry Laudan, Hilary Putnam, and Bas van Fraassen), he is most seriously exercised by the ideas of a particular group of historians and sociologists of science, many of whom are institutionally as well as intellectually affiliated and a number of whom, not insignificantly here, are or have been Kitcher's colleagues at the University of California–San Diego. This group includes Barry Barnes and David Bloor, founders of the "strong programme" in the sociology of science at Edinburgh; their British and American associates, Simon Schaffer, Andrew Pickering, Steven Woolgar, and Steven Shapin (the latter now at San Diego); and, perhaps preeminently, Bruno Latour, the classification-resistant anthropologist, sociologist, and theorist-at-large from Paris who was, for a time, also Kitcher's colleague in California. I'll return to the significance of these institutional overlaps and intersections, but my interest for the moment is the structure of Kitcher's engagement — or, rather, nonengagement — with the ideas of this group.

The question that Kitcher himself sees as most fundamental is whether, in the last analysis, the propositions of science reflect "stimuli from external asocial nature" or if, as some revisionists seem to claim, they are the product of something else distinctly social and verbal, such as (in his words) "social forces," "conversations with peers," or "remarks made by teachers, friends, colleagues, and adversaries."[7] The crucial issue, he writes, is "the constraining power of . . . nature," whether or not, "given the actual social structures present in scientific communities, the input from asocial nature is sufficiently strong to keep consensus practice [i.e., the generation of scientific truth or knowledge as such] on track."[8]

The alternatives posed by Kitcher's formulations here — namely, asocial nature versus mere social exchange, empirical observation versus mere conversation, and the inexorable progress toward truth versus the deflecting pressure of mere exterior forces — are certainly familiar, and so is the structure of distinction and opposition through which he frames them. There is some question, however, as to whether, as he maintains, the crucial issue is the *choice* between those alternatives or, as his adversaries would see it, the coherence of just that notion of choice, of just that set of alternatives, and of the entire system of concepts and conceptual routines marked out by just such familiar but, in their view, dubious distinctions and oppositions. In a revealing footnote, Kitcher declares himself baffled by Latour's rejection of *both* "nature" *and* "society" as explanatory concepts in the history and sociology of science.[9] "I find myself," Kitcher writes, "quite at a loss in understanding what resources are left for understanding the genesis and modification of scientific cognitive states."[10] The sense of perplexity he expresses here — the feeling that something obvious and necessary has been arbitrarily removed, a conceptual space suddenly evacuated, an indispensable resource inexplicably annihilated — is a recurrent and perhaps inevitable result of a collision of ideas of this kind and order.

Read from a post-Legendary perspective, Latour does not, of course, reject Nature or Society per se. Rather, he exposes the instability of the classic dualism that defines and constitutes each of these concepts by mutual contradistinction — or, to put it another way, what he rejects is just the idea of their *per se–ness*. Similarly (and contrary to common charges), Latour and other revisionists do not reject Reason or Rationality per se. What they reject, rather, is the conception of reason as a distinct, ortho-tropic process that can be separated from — or ideally, as in science, purified of — the supposed pressures and distortions of such supposedly exterior forces as the reasoner's individual embodiment, immediate situation, prior intellectual investments, and ongoing verbal interactions. It is clear that part of the difficulty here — as elsewhere in current epistemological controversies — is a crucial divergence of conceptions of *concepts,* especially with regard to their relation to *language.* In the view of Latour and a number of other epistemological revisionists — or, as they are sometimes called in this respect, "constructivists" — all the classic concepts in play in these debates (e.g., "nature," "reason," "reality," and "knowledge") are best understood and treated as, precisely, *constructs,* that is, as variable discursive and conceptual products of our ongoing interactions with the physical, cultural, and verbal worlds in which we live and act. In the view of Kitcher and most traditional realists, however, those classic concepts are properly understood and treated as *autonomous entities,* that is, as the ontologically prior and independently determinate "referents," as it is said, of the words that merely name them.[11] Other important differences of conceptualization are closely related to this one. For example, Kitcher sees what he calls "the effects of nature" as exclusively unidirectional — informational "inputs," as he also calls them — and, in relation to the formation of scientific knowledge, as necessarily prior and causal. For Latour and other constructivists, however, the effects of nature are, precisely, *effects;* that is, what we call Nature can be seen as the relatively stable product of the ongoing reciprocal coordination of our perceptual, conceptual, verbal, manipulative, and other practices, formed and maintained through the very processes of our acting and communicating in the worlds in which we live.

Reciprocal coordination is the key idea here: not social interaction or discourse alone, and not social interaction or discourse simply added to empirical evidence, as the latter is classically understood, but a complex interactive process that is simultaneously dynamic, productive, and self-stabilizing. I shall have more to say about this process later. What is significant for the moment is that the sources of Kitcher's bafflement appear to be more complex than he recognizes and that they reflect a divergence of ideas more radical than he might be willing to grant as possible. But this, of course, is just the issue of incommensurability.

A more general set of observations may be offered at this point. What sustains the recurrent impasses in these and related epistemological controversies are not, I think, just differences, as is sometimes said, of "vocabulary," or conflicts between limited sets of already charted "positions," but rather systematically interrelated divergences of conceptualization that emerge at every level and operate across an entire intellectual domain. The exasperation and sense of intellectual (and sometimes moral) outrage that often attends these failed exchanges can be understood accordingly. Various scholars — historians and sociologists of science, epistemological theorists in related fields, and a number of philosophers as well — have, by one route or another, come to operate conceptually and to interact discursively with their professional and intellectual associates through currently heterodox conceptual idioms. For these scholars, the terms, concepts, and distinctions of traditional epistemology and philosophy of science are no longer either workable or, for the most part, necessary in conducting their professional and intellectual lives. For that reason, they find it usually difficult and sometimes (given the limits of mortal beings) impossible to answer theoretical arguments framed in the traditional terms or appealing to the traditional distinctions and oppositions — impossible, at least, to answer in ways that anybody finds gratifying or dignified. Conversely, traditionalist philosophers of science are, by definition, trained in, committed to, and in a sense intellectually and professionally constituted by a particular epistemological orthodoxy. Accordingly, they operate quite well with the traditional concepts, terms, and distinctions — at least within the orbits of the principal domains of their intellectual lives. And also accordingly, they are likely to find the critiques and alternatives elaborated by their heterodox colleagues absurd, arbitrary, and nihilistic: unmotivated rejections of what is commonsensical, solid, and well-established; irresponsible flattenings-out of what must be, and has been, carefully distinguished; reckless abandonments of what is most desirable and indispensable.

Given the matched and mirrored difficulties just described, it is not surprising that the misconnections in these exchanges are sometimes spectacular. A more extended example is Kitcher's reading and discussion of Latour and Woolgar's book, *Laboratory Life: The Construction of Scientific Facts*, which is an ethnographic study of the day-to-day doings — technical experiments, casual conversations, formal meetings, preparations of scientific articles, and so forth — of a group of scientists in a particular laboratory at the Salk Institute. Explaining an important feature of their own work, Latour and Woolgar observe at one point, "We do not use the notion of reality to account for the stabilisation of a [scientific] statement . . . because this reality is formed as a consequence of this stabilisation."[12] Kitcher cites the remark as evidence of Latour and Woolgar's commitment to the "extreme view that inputs from nature are impotent" or, as he also paraphrases that supposed view, that scientific statements are the product only of the "social arrangement" of a particular laboratory and that "acceptance of [such] statements as firm parts of consensus practice is to be explained in a . . . fashion that makes no reference to the constraining power of stimuli from external, asocial nature."[13] The authors of *Laboratory Life*, Kitcher writes, want us to understand that "the encounters with nature that occurred during the genesis of [the scientists'] belief about TRF [the substance ultimately identified as a particular chemical compound] played no role" in the formation of that belief. In short, he concludes, in apparently aghast italics: "*However those encounters had turned out the end result would have been the same.*"[14]

It could be argued, however, that, contrary to Kitcher's scandalized interpretation of it, the point of the observation he cites from Latour and Woolgar is not that nature is impotent or that reality is infinitely socially malleable, but that to appeal to what the scientific community now accepts as an established fact in order to *explain* how that fact *came to be* established is to *explain* nothing at all: it is only to tell again the familiar (Legendary) story of scientific manifest destiny—that is, the story of how the truth always comes out in the end. But that, of course, is exactly the story that Kitcher himself would tell and seeks most strenuously to defend.[15] Latour and Woolgar's forbearance from present-privileging assumptions is a significant methodological feature of Edinburgh-tutored sociology of science and constructivist science studies more generally. To Kitcher, however, that forbearance looks like gratuitous skepticism: an unnecessary and irrational refusal to credit the truth of established scientific knowledge.[16] It is one of the ironies of the present scene of controversy that just this scrupulous self-skepticism on the part of constructivists—or what, mutatis mutandis, could be called in the classic idiom their "striving for impartiality and objectivity"—is routinely indicted by defenders of traditional realist epistemology as evidence of their reprehensible "relativism."

Kitcher is persuaded by his own perception, interpretation, and report of Latour and Woolgar's ideas that those ideas are absurd. What is significant here is not simply that he misunderstands and misrepresents them, but that, given his paraphrases of their specific claims and arguments in the idiom of his own intellectual tradition and disciplinary culture, he could hardly avoid doing so.[17] Throughout his book, Kitcher employs the idiom of realist/rationalist epistemology—"inputs from asocial nature," reconstructed "reasoning processes," "the constraining power of stimuli" versus "social forces," and so forth—as if its lexicon and syntax (terms, concepts, distinctions, oppositions, and so forth) were altogether unproblematic and, indeed, as if this idiom were the inevitable language of serious thought on questions of knowledge and science. Relevantly enough, he observes good-naturedly at one point that "[f]ew are born antirealists," that such ideas, strongly counterintuitive as Kitcher himself experiences them, can only be the result of a certain line of argument being "thrust upon them."[18] But, of course, few— or none—are born realist/rationalist epistemologists either, however intuitively natural and inevitable that line of argument feels to those who argue it.

Kitcher finds the accounts of science and knowledge offered by constructivist social scientists absurd. He is perplexed, however, not only by the accounts themselves, but also by the fact that they are advanced by people he has reason to think are intellectually competent and indeed highly accomplished: several of them, we recall, including Latour and Shapin, are known to him as colleagues. Accordingly, he ventures a number of explanations for this curious situation. For example, he suggests at one point (perhaps humorously) that Latour's rejection not only of Nature (which might have been expected of a sociologist) but also of Society reflects an "admirable," though clearly misplaced, fondness for "formal symmetry."[19] Or, he remarks at another point, the social-political account of the ascendancy of the experimental method offered by Shapin and Schaffer in their study, *Leviathan and the Air-Pump*, derives from their exaggerated sense of the significance of the theory-ladenness of observation.[20] The latter suggestion bears some scrutiny, for, from a constructivist perspective, the significance of theory-ladenness in intellectual history is hard to exaggerate. Moreover, the phenomenon itself appears to be deeply implicated in the misconnections that concern us here. Both points are vividly illustrated in Kitcher's reading of *Leviathan and the Air-Pump*.

Shapin and Schaffer seek to demonstrate that concerns about the political authority of citizens and sovereigns in seventeenth-century England helped shape the contemporary controversy between Thomas Hobbes and Robert Boyle over the epistemic authority of experimentation versus deduction; and, more significantly and controversially, they suggest that the considerations involved in the political debate were important in determining the *outcome* of the intellectual debate. Commenting on their analysis, Kitcher writes: "Because [Shapin and Schaffer] are so convinced of the power of underdetermination arguments" (another way of stating the idea of theory-ladenness), they fail to focus on "the gritty details of the encounters with nature" and "the complexities of the reasoning about a large mass of observations and experiments"—details and complexities, he suggests, that would, if focused on with "extrem[e] car[e]," turn out to have been decisive in Boyle's victory.[21] As Kitcher acknowledges, he has not himself undertaken this purely hypothetical observation of those purely hypothetical gritty details and complexities of reasoning, nor does he cite any alternative account of them. Nevertheless, so laden is he, so to speak, with his theory of the minor significance of theory-ladenness and the decisive significance of "encounters with nature" and proper "reasoning[s]" that he is prepared to affirm that "when that is done," Shapin and Schaffer's "thesis becomes implausible" and his own view of the history of science is vindicated.

The degree of unabashable conviction displayed in this argument is remarkable, but more significant here is the self-affirming circularity through which it operates. The cognitive process exemplified by Kitcher's resilience in the face of contrary argument and, arguably, contrary evidence is sometimes called the theory-ladenness of observation, sometimes the underdetermination of theory by fact, sometimes the hermeneutic circle, and sometimes the reciprocal determination of perception, belief, and behavior. That process or tendency, which I discuss elsewhere as "cognitive conservatism,"[22] is, I believe, crucial to the dynamics not only of all theoretical controversy but of all theory, which is to say, all knowledge and cognition at both the micro and the macro level. I return to these points below.

In setting forth his own views of how to understand the *history* of science, Kitcher stresses that science is a definitively *epistemic* enterprise, with the single, uniform, and unchanging goal ("independent of field and time") of "attain[ing] significant truth"—which he explains as "charting divisions and recognizing explanatory dependencies in nature."[23] Accordingly, he pays little attention to the historical and ongoing development of laboratory tools, skills, and techniques or to the historical and ongoing emergence of technological applications—all of which he evidently sees as incidental to the macrodynamics of science (or, in his teleological view of those dynamics, to its "advancement") and as irrelevant to the central goal of the *philosophy* of science, which is, in his view, the reconstruction of the reasoning processes of winning arguments.[24] Kitcher's views and practices in these respects have important consequences both for his understanding of revisionist science studies and for his rejection of them.

First, because laboratory techniques, technological applications, and other so-called noncognitive matters are bracketed out in his own conception of science, Kitcher is not disposed to recognize their role in the alternative accounts developed by revisionist historians and sociologists. Thus he seems not to have noticed the important idea, associated with the more recent work of Latour, Pickering, Michel Callon, and others, of the complex relations between laboratory routines, technological extensions, and the formation of scientific statements themselves—all of which are seen in their work as reciprocally motivating, reciprocally determining, and (in Pickering's terms) mutually stabilizing practices.[25] In other words, the dynamics of science are understood in these accounts as neither the dis-covering (removing the covers from) a prior, autonomous truth nor the fabrication (making up whole-cloth) of a sheer collective fantasy but, rather, as the ongoing coordination (weaving together) of observation, theory formation, and material manipulation, each of these being continually adjusted in relation to the others. Thus, details of theory are adjusted to details of technical manipulation and consequent observation; focal points of observation are adjusted to extensions of both technological and theoretical application; material manipulations and details of theory are adjusted to emergent observations; and so on, around again.[26]

The situation of operative conceptual, discursive, and pragmatic stability that emerges from these kinds of ongoing reciprocal coordination is what we often call, in relation to the practices of science, *truth*. In relation to activities of individual cognition or belief formation, the corresponding situation of operative stability is what we usually call *knowledge*.[27] And, in relation to what can be seen as cognition in its broadest sense—that is, an organism's self-maintaining coordination with the domain of its operations[28]—the corresponding situation is often called *adaptation* or biological *fitness*. Of course, truth is commonly attributed not to sets of interactive practices but just to statements; knowledge is commonly seen not as the global state of an organic system but as a specifically mental state; and fitness is treated often enough (as we shall soon see) not as a phenomenological feature of the ongoing interactions of organisms with their environments but as an inherent property of organisms themselves. In each case, what *could* be seen as the name we give to the state of a dynamic system as viewed from a particular perspective is classically or commonly seen as an objective property of a logically and/or ontologically prior, autonomous entity.

These alternative and perhaps rival conceptualizations of truth, knowledge, and fitness are commonly framed *asymmetrically* by the parties on both sides as Our enlightened truth versus Their error or illusion. But the alternatives could also be framed *symmetrically* on both sides as reflections of Our/Their differences of conceptual style and cognitive taste, differences that would themselves be seen as products of Our/Their more or less extensive differences of individual temperament and intellectual history, as played out within more or less different disciplinary cultures and sustained under more or less different epistemic conditions. Of course, these alternative and perhaps rival ways of describing and explaining alternative conceptualizations (either symmetrically or asymmetrically) could *themselves* be described and explained either asymmetrically or symmetrically: either as (for example) Their hopelessly old-fashioned, asymmetrical realist dogma versus Our genuinely enlightened, symmetrical constructivist revelations or (reversing the perspective) as Their trendy, dangerous relativism versus Our established, crucially necessary, normative epistemology — or, again, as differences between Our/Their diversely shaped and situated conceptual styles and cognitive tastes. Thus the linked epistemological issues of — and commonly matched positions on — explanatory a/symmetry and evaluative in/commensurability could reduplicate themselves ad infinitum, at least theoretically. It is worth stressing, however, that they need not — and perhaps never can — do so in practice.

Indeed, it appears that a taste for and commitment to unbroken epistemic symmetry ("relativism" in that sense) on the part of constructivist epistemologists may — and perhaps inevitably does — lapse (or rise) at certain psychologically or rhetorically significant points into a taste for asymmetry and an exhibition of unapologetic epistemic self-privileging. Thus, for example, Latour, distancing himself from an "absolute relativism" that he attributes to some of his science studies associates, stresses that the theoretically presumptive epistemic symmetry (or potentially equal credibility) of rival scientific (or other) accounts is actually always being broken by history and politics: that is, under the prevailing relevant conditions (institutional, intellectual, and so forth), one particular account will be more credible to the relevant populations because it operates with a more powerful and extensive efficacy (of various kinds) than its rivals do. And, he suggests, the *recognition* of this historical and pragmatic asymmetry constitutes, in effect, a more enlightened relativism.[29] Since, as Latour would no doubt grant, the superior efficacy of an account can be determined only after the fact, he would have to be seen as premature in maintaining the epistemic superiority of his own (relative) relativism *now*. On the other hand, the rhetorical energy and power that his account secures at the expense of historical modesty may turn out, in the long run, to be what makes it more effective—and more credible—than the more scrupulously symmetrical or epistemically modest accounts of his rivals, sociological as well as philosophical.

Returning to Kitcher's state of conviction, it appears that, precisely because revisionist accounts of the reciprocal determination of conceptual and material practices in the history of science do not conform to his conception of science as unidirectional progress toward propositional truth, many of the substantive features of those accounts are, in effect, invisible to him. These include, significantly enough, "gritty details" (such as laboratory manipulations, technological extensions, and the effects of material tools and physical skills) that are quite at odds with the all-in-the-head, nothing-but-language, mere-social-forces caricatures of constructivist accounts that alarm traditionalists and are staples of current backlash publications. Because those features of revisionist science studies are invisible to Kitcher, however, they cannot affect either his conception of science or his understanding and evaluation of the revisionist accounts.

The negative route to self-confirming coherence and cognitive immobility indicated by Kitcher's reading practices—where prior theoretical commitment leads to conceptual bracketing-out, which leads to selective perception, which leads to sustained theoretical commitment, and so on, around again—is, of course, just the reverse side of the cognitive mechanism referred to above as the hermeneutic circle, the theory-ladenness of observation, and the reciprocal determination of perception, belief, and behavior. But here we have a very curious and instructive situation. For, as already suggested, the self-securing circularity by which that unhappy—logically objectionable, psychologically embarrassing, cognitively confining—mechanism operates could be seen as duplicating, at the level of individual cognition, the complex processes of reciprocal determination or mutual stabilization that are, according to revisionist science studies, central to the dynamics of scientific practice and, according to revisionist theoretical biology, central to the dynamics of all living systems. What I would emphasize here is not only the evident importance of circularity to everything we call cognition, but also the evidently irreducible *ambivalence* of all the relevant mechanisms or processes, where what is most problematic (circular, self-immuring) duplicates what is most essential (coherence-maintaining, life-sustaining), and where what appears positive from one perspective or at one level of analysis appears negative from another perspective or at another level of analysis — or, in other words, where it is difficult to say, simply or finally, what's good and what's bad.

But this brings us back to Kitcher's book, one of the major goals of which is to affirm both the *normative* mission of traditional philosophy of science — that is, precisely its effort and claim to say what's good and what's bad, to distinguish genuine science from pseudoscience and right thinking from wrong thinking — and its related "meliorative" project, which he explains as "[t]he delineation of formal rules, principles, and . . . informal canons of reasoning, [which,] when supplemented by an appropriate educational regime, can . . . make people more likely to activate propensities and undergo processes that promote cognitive progress."[30]

Kitcher defends the normative ambitions of philosophy of science by way of a parallel to evolutionary biology — under, it must be added, the strongly progressivist and heavily adaptationist interpretation of evolution favored by most realist epistemologists.[31] "Darwinians," he writes, "want not only to claim that successful organisms are those that leave descendents, but also to investigate those characteristics that promote reproductive success"; and analogously, he claims, philosophers of science, reviewing the historical fortunes of various competing scientific theories, want to identify what it is that "confers explanatory and predictive success" on those that succeed.[32] The key to that success, he observes, is clear to the *realist:* just as certain organisms succeed because they are adapted to their environments, certain theories in science succeed "because they fasten on aspects of reality" — which is to say, because they are true.[33] These parallels between traditional, normative philosophy of science and evolutionary biology are, he remarks, "thoroughly Darwinian," the emphasis being required, perhaps, because significantly different interpretations of evolutionary theory could be offered and, as Kitcher acknowledges, significantly different analogies (and lessons) have in fact been drawn. It could be observed, for example, that since the biological fitness of an organism can be specified only in relation to a particular, contingent, environmental situation, the idea of *general* "sources of fitness" makes little sense and the search for *inherent* traits that "endow organisms with high Darwinian fitness" or "dispose them to survive" is pointless.[34] Analogously, we are no more able to devise a method for distinguishing or producing "cognitively progressive," or *inherently* more likely to be true, beliefs than to devise one for distinguishing or producing *inherently* more likely to be fit organisms . . .

or people. Indeed, if Darwinian "fitness" is taken as a metaphor for epistemic truth, then the meliorative project of traditional epistemology would have to be seen as the eugenics of philosophy.

We may now turn to the ways in which right and wrong thinking are actually separated in Kitcher's account of rationality. Insisting, as always, on the supposedly "relativism"-dashing availability of objective criteria for the assessment of cognitive activities and products, he writes: "People can make cognitive mistakes, perceiving badly, inferring hastily, failing to act to obtain inputs from nature that would guide them to improved cognitive states. . . . Some types of processes are conducive to cognitive progress; others are not."[35] To illustrate the difference between these types, he contrasts the ("reconstructed") reasoning of Darwin and his followers with the cognitive intransigence of nineteenth-century skeptics and present-day creationists. Reflecting on the latter's current debate with Darwinians (a debate, it should be stressed, in which Kitcher himself has participated extensively[36]), he writes:

> The behavior of creation scientists indicates a kind of inflexibility, deafness, or blindness. They make an objection to some facet of evolutionary biology. Darwin's defenders respond by suggesting that the objection is misformulated, that it does not attack what Darwinists claim, that it rests on false assumptions, or that it is logically fallacious. How do creation scientists reply? Typically, *by reiterating the argument.* Anyone who has followed exchanges in this controversy . . . sees that there is no adaptation to any of the principal criticisms.[37]

He means, of course, that there is no adaptation by creationists to the criticisms of their views by Darwinists such as himself;[38] but the "anyone" who sees this could not be quite anyone, since creationists could observe that, as far as adapting to criticism goes, Darwinists—blind, deaf, and inflexible as anyone can see they are—have not budged an inch either. Kitcher explains the overt intransigence of creationists as a sign of their underlying cognitive unwholesomeness. Creationists, however, could probably give a comparable array of reasons for their opponents' stubbornness in error: ignorance of the Bible, secular humanist prejudice, modern infatuation with evolutionary theory plus, perhaps, certain sins of sloth and pride. My point here is not, of course, that the opinions of the Darwinists and the creationists regarding evolutionary theory are "equally valid," but that, for all the differences in their favored idioms and authorities, the explanation *Kitcher* offers for the cognitive intransigence of his longtime adversaries exhibits the same asymmetrical structure as *their* explanation for *his*, which is to say the same, perhaps endemic, tendency toward absolute epistemic self-privileging.[39]

Kitcher claims that the distinctions he draws between proper and improper activations of good and bad cognitive propensities are based on objective norms and criteria. There is reason to suspect, however, that here — as commonly elsewhere in the case of such objectivist claims — the judgments of goodness and badness, propriety and impropriety, *preceded* and were indeed *presupposed* by the framing of those norms and criteria. Latour remarks, in a passage that Kitcher quotes with some exasperation, that all efforts to separate rational and irrational belief are no more than *"compliments or curses,"* saying nothing about the beliefs in question but "simply help[ing] people to further their arguments as swear words help workmen to push a heavy load, or as war cries help karate fighters intimidate their opponents."[40] Kitcher objects that such remarks "disguise both the serious purpose and the genuine difficulties involved in appraisals of rationality."[41] But perhaps we have here just another (disguised) curse or compliment. For how, and from what presumptively objective perspective, can it be determined whether Latour is disguising a serious purpose and genuine difficulties or exposing an earnest but vacuous enterprise?

To draw together, now, a number of points touched on above: First, in connection with Kitcher's distinctly asymmetrical and often overtly self-privileging notions of mental fitness, we may recall that cognitive conservatism — the process or mechanism that produces what we call, under some conditions, circularity and stubbornness and, under other conditions, coherence and stability — is conceived here not as an inherent flaw in certain (*other*) people's cognitive design but as an endemic tendency or characteristic of human (and perhaps not only human) cognitive operations.[42] Indeed, it appears that hermeneutic circularity, the theory-ladenness of observation, and, more generally, the reciprocal determination of belief, perception, and behavior are crucial features of that complex set of cognitive processes we call — depending on where we are standing and how we are cutting it — perception, reasoning, thinking, belief formation, theory formation, experiencing, responding, behaving, or living. The value — "fitness," "functionality," "progressiveness," "success" — of those processes cannot be indicated or characterized independently of the domains in which they are played out or of the perspectives from which their products (i.e., particular beliefs and related behaviors) are assessed. The cognitive processes that, on occasion, lead us (or is it only *them?*) astray and confine our thinking to circles of self-confirming self-affirmation appear to be the very *same* processes that give coherence to our individual beliefs, that sustain and stabilize all scientific knowledge as such, and that lead us to what we sometimes call *truth.*[43]

Second, foregrounding the idea of cognitive or epistemic *domains* — that is, the spaces or, one could say, niches in which we play out our particular beliefs — we may recall here the institutional overlaps and interconnections noted earlier among the participants in these controversies. What can be stressed now is that the academic and professional arenas in which the various parties play out their more or less divergent ideas may themselves diverge or coincide to greater or lesser degrees. Philosophers, historians, and sociologists of science, respectively, typically belong to distinct disciplinary cultures, publish in different professional journals, and train different graduate students. In these respects, their epistemic domains are relatively discrete. At the same time, however, they may be located in the same universities, attend some of the same interdisciplinary conferences, teach some of the same undergraduate students, and write for some of the same general interest magazines. In these respects, their epistemic domains will overlap, and they — and their respective beliefs — will inevitably, and for better or worse, bump into each other. Where the domains in question are relatively discrete, there is little occasion for the *divergences* of their beliefs to become *conflicts,* and their respective ideas and idioms can continue, so to speak, to live side by side. It is, of course, where those domains overlap or coincide that divergences of belief and conceptual idiom, and related differences of cognitive taste and disciplinary projects and practices, do become conflicts, as exhibited, for example, in the debates I have examined here and, in some places, in active, sometimes bitter, rivalries for intellectual and institutional authority.

Epistemic authority is involved in other ways as well in these quite general social and cognitive dynamics. Theoretical accounts that are more or less incompatible with what we already take for granted as obvious, self-evident, or unquestionable are likely to appear inadequate, incredible, or incoherent to us, and also, depending on our sense of the intellectual authority and sometimes other social characteristics of the people who offer them — e.g., their institutional credentials, age, gender, or class [44] — as ignorant, silly, outlandish, wildly radical, or fraudulent. We may resist such alternative accounts even though they are presented with detailed arguments and evidence that other people seem to find coherent and compelling; for, of course, those other people may, for that very reason, appear ignorant or intellectually inept to us and/or, depending again on our sense of them otherwise, gullible, trendy, brainwashed, or ideologically motivated. The *energy* with which we resist such accounts will correspond, of course, to the significance to us of the ideas with which they are in conflict. The *form* of that resistance, however, is likely to be shaped by the type and degree of our own intellectual authority in the relevant epistemic domains and may range, accordingly, from perplexed and resentful withdrawal to elaborate condescension, detailed counterargument, virulent attack, or attempted suppression.

Since broader political resonances are inevitable here, a further general point can be added. In situations of intellectual rivalry, it sometimes happens that the only acceptable outcome for at least one of the parties is *absolute epistemic supremacy*: the claim is made, in other words, that there is but one truth, that the party in question is enlightened as to its true nature and proper pursuit, and that it is universally desirable that all this be universally acknowledged. In such cases, any divergence of professed belief, conceptual idiom, or discursive practice in any domain whatsoever is seen as dangerous error requiring intervention and correction — or, in other words, as heresy. Accordingly, all intellectual divergence is seen as deviation, all deviation becomes conflict, and, for the party (or parties: it may be both) so disposed, all conflict becomes zero-sum rivalry, with winners properly taking all, and taking it for all time, and losers properly disappearing forever. Indeed, it is precisely when institutionalized systems of ideas and related conceptual idioms and discursive practices claim absolute epistemic supremacy — or, of course (though there is often no difference), when they entail visions of universal *political* supremacy — that "wars of truth" become duels to the not always figurative death.

My description here is meant to be quite general, but it is not irrelevant that Legend insists on the unity not only of truth but also of epistemic domains. These indeed are its defining orthodoxies in relation to the contra-defining heresies of what it calls "relativism" and, accordingly, constitute a major source of the resistance by traditionalists to the idea of incommensurability and to the related notion of multiple "worlds" — which could also be understood as multiple epistemic domains.

Returning to the issue of incommensurability as framed at the beginning, we may ask where we stand at this point. Having lined up and compared these divergent accounts of science, are we now prepared to choose the better — that is, the epistemically superior — one? From the perspective of this analysis, the question is unanswerable and the choice irrelevant. This is not to say that we cannot or should not assess different ideas, theories, or beliefs; on the contrary, we can and must assess them continuously, in the very process of playing and living them out in the relevant domains of our lives — intellectual, political, technological, religious, and so forth. It is to say, rather, that the occasions for such terminally decisive adjudications, as they are classically depicted, never arise.

We may note here Kitcher's somewhat vehement insistence on "the limits of proper tolerance."[45] One may agree that there are indeed such limits, but observe, in accord with this analysis, that they are mundane — practical and quotidian — matters of social and institutional geography and politics, not matters for the high courts of epistemic adjudication. The point is illustrated well enough by the situation that evoked Kitcher's vehemence: the debate between creationists and Darwinists. As long as the domains in which their alternate accounts (of the origin of species, the mutability of life-forms, the age of the universe, etc.) are played out remain effectively discrete, there is no reason for intellectual or political tolerance to be limited nor, in fact, any occasion for it to be displayed — except, of course, as forbearance from invasive missionary activity by one side or the other. Conflict arises, however, when there is, or threatens to be, a coincidence of domains, as in the demand by some citizens that scriptural accounts of the relevant phenomena be taught in American public schools in place of, or as an "equal time" alternative to, evolutionary theory. It is quite a temptation but, from the present perspective, a conceptual mistake for Darwinian-minded citizens to imagine this conflict on the model of the struggle between Galileo and the pope or between Darwin himself and his nineteenth-century clerical adversaries. It is certainly a *strategic* mistake for them to play it that way at local school board meetings or in the nation's courts. For, unless Darwinists agree to having the issue framed in such terms, the relevant question is not whether evolutionary theory satisfies such arbitrary and arguably vacuous *general* epistemic criteria as "incontrovertible factuality" but, rather, which authorizing institutions are appropriate for evaluating the material to be taught specifically in American public schools. Given the Constitutional separation of church and state, it could be argued that, although scriptural and other religious authority is appropriate enough in parochial schools, the only appropriate institutional authorities for assessing public school materials are secular. That would mean, in this case, that any theory of the origin of species, the mutability of life-forms, or the age of the universe (etc.) taught in science classes in American public schools is properly assessed in relation to currently established scientific knowledge and practice, where "established" is understood as *broadly accepted by members of the relevantly authorized, secular epistemic communities.* Alternative interpretations of "established" as "incontrovertibly factual" or "determined as finally, objectively, and transcendentally true" could be seen, accordingly, as red herrings. Red herrings can be rhetorically effective, of course, at school board meetings and even in courtrooms. But the effectiveness of this one has depended, it appears, on the readiness of some Darwinists (including some biologists and philosophers of science) to rise to the epistemic-supremacist bait dangled by opponents.[46]

There are, it appears, few particular occasions and no particular ways to select winners and losers in theoretical controversies, and, in a sense, no winners or losers either, at least no objectively determinate ones. It seems clear that none of the presently rival orthodox/heterodox accounts of science will survive or endure in any of their present forms. There will be "advancements" and retreats, of course, but only in the sense of the increased or decreased authority of one or another such account in various more or less restricted epistemic niches. In other words, the fitness, success, or survival of any theoretical account of science will still be measurable only in relation to particular conditions and only from particular perspectives. And the same could be said of any theoretical account whatsoever.

There can be no ultimate comparison of or decision between the epistemic merits of these rival theories, moreover, because each of them is being transformed by, among other things, the dynamics of their very rivalry. Academic philosophy of science has undergone substantial transformation at both the individual and the institutional level since at least the 1960s, when Kuhn and Feyerabend first presented their unignorable challenges to Kitcher's Legend. Claims, methods, and missions have been modified, in some respects drastically. Alliances have been formed with other disciplines, including the biological and physical sciences themselves, and new interdisciplinary fields, such as cognitive science, have emerged and become relatively well established.[47] The reciprocal of this is also occurring. Revisionist science studies keeps revising itself in response to, among other things, the resistances of traditional realist/rationalist epistemology. To mention only one example here, but a telling one: Pickering's increased emphasis on "material practices" in his accounts of the history of particle physics and his related delineation and embracing of a position he calls "pragmatic realism" appear to have been shaped by, among other things, his prior and ongoing interactions with some persistently resistant philosophers of science.[47]

Incommensurability is, it appears, neither a logically scandalous relation between theories, nor an ontologically immutable relation between isolated systems of thought, nor a morally unhappy relation between sets of people, but a contingent experiential relation between historically and institutionally situated conceptual and discursive practices. Some radically divergent ideas never meet at all, at least not in the experience of mortal beings. In other cases, meetings are staged repeatedly but never come off, ending only in mutual invisibility and inaudibility. Sometimes, however, meetings do occur, perhaps intensely conflictual and abrasive but also, in the long run, mutually transformative. Thus it may be that, at the end, on the real Judgment Day — if there is one — for which the philosophers are always preparing us, when all the stories are told and all the chips are in, counted, and compared, we will not only be unable to say who finally won but even to tell which was which.

NOTES

This essay is adapted from Barbara Herrnstein Smith, *Belief and Resistance: Dynamics of Contemporary Intellectual Controversy* (Cambridge, MA, 1997).

1   Philip Kitcher, *The Advancement of Science: Science without Legend, Objectivity without Illusions* (Oxford, 1993).

2   Allusions to Scylla and Charybdis are common in contemporary theoretical controversy, along with more general statements as to the desirability of steering a middle course between such alleged extremes as objectivism and relativism, realism and constructivism, old-fashioned rationalism and newfangled postmodern irrationalism, and so forth. The advertised via media usually turns out to be, as here, a (con)temporized version of received (e.g., objectivist, realist, rationalist) wisdom.

3   Kitcher, *Advancement of Science*, 127.

4   Ibid., 178–218.

5   See, for example, J. A. Kegley's review in *Choice* (November 1993): 471–72; and David Papineau, "How to Think about Science," *New York Times Book Review*, July 25, 1993, 14–15.

6   See, for example, John Ziman, "Progressive Knowledge," *Nature* 364 (1993): 295–96; and Steve Fuller, "Mortgaging the Farm to Save the (Sacred) Cow," *Studies in the History and Philosophy of Science* 25 (1994): 251–61.

7   Kitcher, *Advancement of Science*, 166, 162, 164.

8   Ibid., 166, 165.

9   See Bruno Latour, *The Pasteurization of France*, trans. Alan Sheridan and John Law

(Cambridge, MA, 1988); and *We Have Never Been Modern*, trans. Catherine Porter (Cambridge, MA, 1993).

10   Kitcher, *Advancement of Science*, 166 n. 52.

11   Kitcher seems unaware of critiques of the Fregean referentialist model of language to which he appeals or of related alternative accounts. Given the mutual segregation of Continental and Anglo-American philosophy, it is not surprising that neither Foucault nor Derrida appears in his lengthy bibliography. It is surprising, however, that neither Wittgenstein nor Rorty — or, aside from one minor brush-off, Hesse — does. For related discussion, see Michael A. Arbib and Mary B. Hesse, *The Construction of Reality* (Cambridge, 1986), 147–70.

12   Bruno Latour and Steven Woolgar, *Laboratory Life: The Social Construction of Scientific Facts* (Princeton, 1986 [1979]), 180.

13   Kitcher, *Advancement of Science*, 164, 167, 165–66.

14   Ibid., 166.

15   With a glance, it seems, at S. J. Gould and R. C. Lewontin's celebrated essay, "The Spandrels of San Marco and the Panglossian Paradigm: A Critique of the Adaptationist Programme," *Proceedings of the Royal Society of London* 205 (1978): 581–98, Kitcher explicitly rejects the telling of "just-so-stories" in the history of science. Nevertheless, he cites, endorses, and is evidently influenced by Howard Margolis, *Paradigms and Barriers: How Habits of Mind Govern Scientific Beliefs* (Chicago, 1993), which features a defiantly and explicitly Whiggish history of science. Margolis maintains that the symmetry postulate of the Edinburgh sociologists and historians of science — that is, their refusal to privilege present scientific knowledge methodologically as always already true — is a foolish overreaction to "an older history of science" that got "a bad reputation" because it said impolite things about the losers in scientific controversies. That was, Margolis observes, crude — but, he adds, "of course, being winners, the winning side must have had more of *something*" (197; his emphasis). I take up Kitcher's version of that "something" below.

16   Kitcher, *Advancement of Science*, 188.

17   Misunderstandings and misrepresentations occur on both sides of these debates, of course, for revisionists as well as traditionalists interpret the arguments of their adversaries through their own assumptions. Symmetry-conscious constructivist epistemologists such as Latour and Woolgar would presumably acknowledge this in principle but, since one is always blind to one's own blind spots, would not be able to point out their own misunderstandings and misrepresentations — nor, given my own perspective on these issues, am I well situated to do so for them.

18   Kitcher, *Advancement of Science*, 131.

19   Ibid., 166 n. 52.

20   Steven Shapin and Simon Schaffer, *Leviathan and the Air-Pump: Hobbes, Boyle, and the Experimental Life* (Princeton, 1985). Kitcher believes that a "pessimistic" (156) overestimation of the significance of theory-ladenness is a general feature of contemporary sociology of science. The claim that "we see just what our theoretical commitments would lead us to expect" is "a gross hyperextension of what philosophers and psychologists are able to show" (*Advancement of Science*, 167 n. 53; see also 141 n. 18). His statement of that claim is itself something of a hyperextension, however, setting up a spurious contrast with the "eminently sensible conclusion," attributed to Kuhn, that "anomalies emerge in the course of normal science." The crucial

issue, of course, is not the emergence of anomalies—something that no sociologist of science would, I think, deny—but how to describe their operation in intellectual history. Kitcher evidently sees them as epistemic arrows shot straight from reality, piercing our otherwise theory-clouded or theory-skewed observations and setting us, and our theories, straight. The alternative view—and, arguably, the one Kuhn himself favors—is that perceived anomalies may destabilize specific theories but, like all other perceptions, must themselves be interpreted via prior conceptualizations. For detailed discussion of Kuhn's views on this and related topics, see Paul Hoyningen-Huene, *Reconstructing Scientific Revolutions: Thomas S. Kuhn's Philosophy of Science*, trans. Alexander T. Levine (Chicago, 1993), 223–44.

21   Kitcher, *Advancement of Science*, 169 n. 55. Theories are "underdetermined" by evidence or observation of fact, and, conversely, observation of supposed fact is overdetermined by, or "laden" with, prior theory.

22   In *Belief and Resistance*, I treat knowledge, cognition, and, by implication, "theory" as embodied processes, practices, and products, not as matters of disembodied or purely formal intellectual activity.

23   Kitcher, *Advancement of Science*, 157.

24   Kitcher defends this traditional goal of philosophy of science rather awkwardly and equivocally. On the one hand, he maintains that the abstractions and idealizations of "philosophical reflections about science"—like the models of economic theorists vis-à-vis "the complicated and messy world of transactions of work, money, and goods"—are necessary to "lay bare large and important features of the phenomena" and to "recogniz[e] the general features of the . . . enterprise." On the other hand, he cautions that "to rebut . . . charges [of unrealistic irrelevancy]—*or* to concede them *and* to do better service to philosophy's legitimate normative project—we need to idealize the phenomena *but* to include in our treatment the features [e.g., the complicated and messy ones?] that critics emphasize" (10; my emphases).

25   See, for example, Bruno Latour, *Science in Action: How to Follow Scientists and Engineers through Society* (Cambridge, MA, 1987); *Pasteurization of France*, and "On Technical Mediation—Philosophy, Sociology, Genealogy," *Common Knowledge* 3 (fall 1994): 29–64; Michel Callon, "Society in the Making: The Study of Technology as a Tool for Sociological Analysis," in *The Social Construction of Technological Systems: New Directions in the Sociology and History of Technology*, ed. Wiebe E. Bijker, Thomas P. Hughes, and Trevor Pinch (Cambridge, MA, 1987), 83–103; Andrew Pickering, "From Science as Knowledge to Science as Practice," in *Science as Practice and Culture*, ed. Andrew Pickering (Chicago, 1992), 1–28; and Wiebe Bijker and John Law, eds., *Constructing Networks and Systems* (Cambridge, MA, 1994). On the collaborative use of instruments in oceanographic research, see also Charles Goodwin, "Seeing in Depth," *Social Studies of Science* 25 (1995): 237–74.

26   Contrary to common misunderstandings, "around again," in both the classic idea of the hermeneutic circle and more recent analyses of the reciprocal determination of theory, action, and observation, describes not a continuous repetition of the same path but—if spatial images are sought—a set of continuously linked loops. For related discussion, see Arbib and Hesse, *Construction of Reality*, 8; and Barbara Herrnstein Smith, "Belief and Resistance: A Symmetrical Account," *Critical Inquiry* 18 (1991): 125–39.

27   For an influential, though not unproblematic, account of stabilization at this level of analysis, see Jean Piaget, *Biology and Knowledge: An Essay on the Relations be-

tween Organic Regulations and Cognitive Processes, trans. Beatrix Walsh (Chicago, 1971 [1967]).

28 See Humberto Maturana and Francisco Varela, *Autopoiesis and Cognition: The Realization of Living* (Boston, 1980), and *The Tree of Knowledge: The Biological Roots of Human Understanding* (Boston, 1988); see also Humberto Maturana, "The Origin of Species by Means of Natural Drift, or Lineage Diversification through the Conservation and Change of Ontogenic Phenotypes," unpublished trans. Cristina Magro and Julie Tetel; published in Spanish as Occasional Publications of the National Museum of Natural History, Santiago, Chile, 43 (1992).

29 See Latour, *We Have Never Been Modern,* 111–14.

30 Kitcher, *Advancement of Science,* 186.

31 For examples, see Ruth Millikan, *Language, Thought, and Other Biological Categories: New Foundations for Realism* (Cambridge, MA, 1984); *Evolutionary Epistemology, Rationality, and the Sociology of Knowledge,* ed. Gerard Radnitsky and W. W. Bartley (La Salle, IL, 1987); and William G. Lycan, *Judgement and Justification* (Cambridge, 1988).

32 Kitcher, *Advancement of Science,* 155–56, 156–57.

33 Ibid., 156. Kitcher illustrates the point with the success of genetics, which he explains as follows: given the crucial role of "references" to "genes" in genetics' explanatory "schemata," the reason why it can explain and predict biological phenomena so well is "that there are genes" (157).

34 Ibid., 156.

35 Ibid., 185–86.

36 See Philip Kitcher, *Abusing Science: The Case against Creationism* (Cambridge, MA, 1982).

37 Kitcher, *Advancement of Science,* 195; his emphasis.

38 Kitcher's term is "Darwinists," not "Darwinians," here, perhaps to indicate that the defenders of evolutionary accounts are not always professional biologists.

39 Interestingly enough, one of Kitcher's most rhetorically effective arguments in *Abusing Science* is, in effect, the epistemic asymmetry of creation scientists, who, he points out, appeal to different criteria for determining the scientificity of the Darwinian account and their own.

40 Latour, *Science in Action,* 192; his emphasis.

41 Kitcher, *Advancement of Science,* 185.

42 The exposure by feminist theorists, among others, of the self-privileging biases commonly involved in normative invocations of supposed human universals have gone some distance toward making any reference to general human traits intellectually and ideologically suspect. I would stress, therefore, that my references here to apparently endemic cognitive processes are *not* normative and that they operate in the argument in a pointedly *anti-*(self-)privileging way.

43 For discussion of related ambivalent and apparently endemic cognitive processes, see R. E. Nisbett and L. Ross, *Human Inference: Strategies and Shortcomings of Social Judgment* (Englewood Cliffs, NJ, 1980); and *Judgment under Uncertainty: Heuristics and Biases,* ed. D. Kahneman, P. Slovic, and A. Tversky (Cambridge, 1982). For a survey of *pseudodoxia epidemica* from a staunchly realist/rationalist perspective, see Massimo Piattelli-Palmarini, *Inevitable Illusions: How Mistakes of Reason Rule Our Minds,* trans. Massimo Piattelli-Palmarini and Keith Botsford (New York, 1994).

44 On the relation between class and epistemic authority, see Steven Shapin, *A Social History of Truth: Civility and Science in Seventeenth-Century England* (Chicago, 1994).

45   Kitcher, *Advancement of Science*, 196.

46   Difficult and important practical situations of this kind—dealing with published denials of the Nazi Holocaust is another example—are often cited as real-life refutations of (supposedly merely theoretical) epistemological relativism. The implication is that at the "limits of tolerance" posed by such situations, the choice can only be between, on the one hand, declaring certain people objective fools or absolute liars and, on the other hand, capitulating to their demands or agreeing to the "equal validity" of their claims. As just indicated, however, and as I have discussed elsewhere, these are not the only alternatives, nor is it clear that the recommended beyond-tolerance responses—that is, the issuing of strenuous absolutist/objectivist declarations of morality and truth—would (in themselves) have (only) the presumably desired outcomes; cf. Barbara Herrnstein Smith, "The Unquiet Judge: Activism without Objectivism in Law and Politics," *Annals of Scholarship* 9 (1992): 111–33; and "Making (Up) the Truth: Constructivist Contributions," *University of Toronto Quarterly* 61 (1992): 422–29.

47   Philosophy of science appears, in some places, to be merging with its own subject sciences, for example, philosophy of biology with theoretical biology; see Robert N. Brandon, *Adaptation and Environment* (Princeton, 1990); and *Conceptual Issues in Evolutionary Biology*, 2d ed., ed. Elliott Sober (Cambridge, MA, 1994). In other places, it seems to have naturalized not only its lingo but also its projects and methods, either jumping ship altogether in anticipation of neurophysiological replacements of philosophical accounts of cognition (see Patricia Churchland, *Neurophilosophy: Toward a Unified Science of the Mind/Brain* (Cambridge, MA, 1986), or reconceiving its task as that of mediating or "intertranslating" the discourses of traditional epistemology and contemporary cognitive science; see Andy Clark, *Microcognition: Philosophy, Cognitive Science, and Parallel Distributed Processing* (Cambridge, MA, 1991) and *Associative Engines: Connectionism, Concepts, and Representational Changes* (Cambridge, MA, 1993); Daniel Dennett, *Consciousness Explained* (Boston, 1991); Owen Flanagan, *Science of the Mind* (Cambridge, MA, 1992); and *Consciousness Reconsidered* (Cambridge, MA, 1992).

48   See Andrew Pickering, *The Mangle of Practice: Time, Agency, and Science* (Chicago, 1995). See also Shapin's recent (and, from the present perspective, dubious) affirmation of the "incorrigible presupposition" of a realist ontology by "virtually any form of praxis" (*Social History of Truth*, 29ff., 122). In his formal acknowledgments there, Shapin alludes to "a series of friendly arguments with my colleague Philip Kitcher" (iii). Kitcher, in turn, notes in the formal acknowledgments of *The Advancement of Science* that his "thinking about epistemology and the history and philosophy of science has been greatly helped by discussions with [among others] . . . Steven Shapin," who, he adds, is not "likely to agree with the conclusions of this book, but can pride [himself] on having diverted me from even sillier things that I might have said" (viii).

Notes on Contributors

MALCOLM ASHMORE, Lecturer in Sociology at Loughborough University, UK, is the author of *The Reflexive Thesis: Wrighting Sociology of Scientific Knowledge* (1989) and coauthor (with Michael Mulkay and Trevor Pinch) of *Health and Efficiency: A Sociology of Health Economics* (1989). He is completing a new book on debunking fraud and error in science.

MICHEL CALLON, Professor of Sociology at the Ecole Nationale Supérieure des Mines, Paris, has published articles on the sociology of science and technology and the economics of research and development. He coedited *Mapping the Dynamics of Science and Technology* (1986) and is currently editing a volume on the construction of markets and writing a book on the modalities of coordination.

OWEN FLANAGAN is Professor of Philosophy and Psychology (Experimental) and Chair of the Department of Philosophy, Duke University. His most recent book is *Consciousness Reconsidered* (1992).

JOHN LAW, Professor of Sociology at Keele University, UK, is the author of *Organizing Modernity* (1994). Together with Michel Callon and Bruno Latour, among others, he has contributed to the development of actor-network theory.

SUSAN OYAMA is Professor of Psychology at John Jay College and the CUNY Graduate School and University Center. The author of *The Ontogeny of Information: Developmental Systems and Evolution* (1985) as well as coauthor and coeditor of *Aggression: The Myth of the Beast Within* (1988), she has written on the nature/nurture opposition, biology and ethics, and the relationship between development and evolution.

ANDREW PICKERING is Professor of Sociology and a member of the Unit for Criticism and Interpretive Theory, University of Illinois at Urbana-Champaign. The author of *Constructing Quarks: A Sociological History of Particle Physics* (1984) and editor of *Science as Practice and Culture* (1992), his most recent book is *The Mangle of Practice: Time, Agency, and Science* (Chicago, 1995).

ARKADY PLOTNITSKY, currently Visiting Scholar at Duke University's Center for Interdisciplinary Studies in Science and Cultural Theory, holds degrees in mathematics and literature from the University of St. Petersburg and the University of Pennsylvania. His most recent books include *Reconfigurations: Critical Theory and General Economy* (1993) and *Complementarity: Anti-Epistemology after Bohr and Derrida* (1994).

BRIAN ROTMAN, an independent scholar who lives with his wife and two daughters in Memphis, Tennessee, has been awarded fellowships by the ACLS, NEH, and Stanford University Humanities Center. He is the author of *Signifying Nothing: The Semiotics of Zero* (1993) and *Ad Infinitum . . . The Ghost in Turing's Machine: Taking God Out of Mathematics and Putting the Body Back In* (1993).

BARBARA HERRNSTEIN SMITH is Braxton Craven Professor of Comparative Literature and English and Director of the Center for Interdisciplinary Studies in Science and Cultural Theory at Duke University. Her most recent books are *Contingencies of Value: Alternative Perspectives for Critical Theory* (1988) and *Belief and Resistance: Dynamics of Contemporary Intellectual Controversy* (1997).

JOHN VIGNAUX SMYTH is chair of the English Department at Northeast Louisiana University. The author of *A Question of Eros: Irony in Sterne, Kierkegaard and Barthes* (1986), he has recently completed a new book, *The Habit of Lying (Or Fundaments of Fiction)*, forthcoming from Duke University Press.

E. ROY WEINTRAUB is Professor of Economics at Duke University. Trained in mathematics, his recent work has explored the interconnection between the history of mathematics and the history of economics. He is the author of *Stabilizing Dynamics: Constructing Economic Knowledge* (1991) and the editor of *Towards a History of Game Theory* (1992).

Index

Actor-network theory (ANT), 96, 102, 113–14, 115n.3
Adaptation, 252
Agency, 6–8, 14, 40–45, 50–67, 73, 75n.1, 76n.9, 83–86, 88n.5, 91–93, 95–114, 116nn.20, 26; arbitrariness, 112; attribution, 113, 116n.26; and causality, 7; conceptual, 7, 40, 42; "dance of agency," 42–43, 61; decentering, 92, 107, 113; disciplinary, discipline in, 41, 43, 45, 50–67, 73, 75n.5, 76n.9, 85, 88n.5, 92; and emergence, 98; in Hamilton, 50–64, 73; human and nonhuman, 7, 9, 61, 95–114; machines and instruments in, 40, 64–65; material, 7, 40, 62, 64–65; in mathematics, 8; and models/modeling, 42, 50–64; multiplicity and heterogeneity, 85, 113; and politics, 113; and structure, 106; and subject/subjectivity, 7
Algebra, 41, 45–49, 58, 61–64, 67–73, 87, 239n.9; foundations of, 45; and geometry, 45, 48, 58, 61–64, 67, 72–73, 170n.44, 239n.9; as language, 72; symbolic, 67, 72, 87; and temporality, 68, 72–73, 82n.75, 87
Alterity, 135–36, 140, 149, 151–53, 167; and efficacy, 151; and materiality, 151; radical (vs. absolute), 135–36, 140–41, 149, 151–53, 167; and reality, 151; reciprocal, 141
Anaxagoras, 135
Angier, Natalie, 123, 129, 131n.15, 132n.16
Anomaly in science, 263–64n.20

ANT. *See* Actor-network theory
Aristotle, 164, 212
Arithmetic, 216–17, 220, 227
Arnold, V. A., 28
Arrow Kenneth, 14, 174–75, 181–85, 188nn.3, 4, 8–13
Ashmore, Malcolm, 6, 9–10, 15, 112, 115nn.14–15, 116n.16, 117n.31, 207n.1, 210n.34
Auster, Paul, 97, 115n.8
Axiom of choice, 236

Baboon experiment, 97–98, 115n.7
Bachelard, Gaston, 76n.11
Barnes, Barry, 9, 75n.6, 80n.53, 81n.68, 116n.24, 245
Barone, Enrico, 183
Bataille, Georges, 136, 151
Bateson, Gregory, 28
Beckett, Samuel, 1, 4, 10, 212–15, 217, 219, 221, 225, 227, 229–34, 237, 238nn.5, 8, 240n.32, 241n.55, 242nn.58, 60–64, 67
Becoming, process, 14. *See also* Emergence
Bell, J. S., 159
Benedikt, Michael, 33, 39n.10
Berkeley, George, 240n.23
Bigelow, John, 215–16, 224, 239n.10, 240n.23
Biology, 1–2, 7–8, 14, 169n.26, 244, 254–55, 266n.46; complexity in, 124–26; complexity and simplicity in, 124–25; evolutionary, 244, 255–56; molecular, 169n.26; revisionist, 254
Biquaternions, 73–74

Library of Congress Cataloging-in-Publication Data
Mathematics, science, and postclassical theory / Barbara Herrnstein
Smith and Arkady Plotnitsky, editors.
The text of this book originally was published without the essay
"Microdynamics of Incommensurability" by Barbara Herrnstein Smith
and without the index as volume 94, number 2 of the South Atlantic
Quarterly — T.p. verso.
Includes index.
ISBN 0-8223-1857-1 (cloth : alk. paper). — ISBN 0-8223-1863-6
(pbk. : alk. paper)
1. Science.   2. Mathematics.   I. Smith, Barbara Herrnstein.
II. Plotnitsky, Arkady.
Q158.5.M368   1996   500 — dc20   96-13178   CIP